Twentieth Century Land Settlement Schemes

Land settlement schemes, sponsored by national governments and businesses, such as the Ford Corporation and the Hudson's Bay Company, took place in locations as diverse as the Canadian Prairies, the Dutch polders and the Amazonian rainforests. This novel contribution evaluates a diverse range of these initiatives.

By 1900, any land that remained available for agricultural settlement was likely to be far from the settlers' homes and located in challenging physical environments. Over the course of the twentieth century, governments, corporations and frequently desperate individuals sought out new places to settle across the globe, from Alberta to Papua New Guinea. This book offers vivid accounts of the difficulties faced by many of these settlers, including the experiences of East European Jewish refugees, New Zealand soldier settlers and urban families from Yorkshire.

The book considers how and why these settlement schemes succeeded, found other pathways to sustainability or succumbed to failure and even oblivion. In doing so, it indicates pathways for the achievement of more economically, socially and environmentally sustainable forms of human settlement in marginal areas. This engaging contribution will be of interest to individuals in the fields of historical geography, environmental history and development studies.

Roy Jones is an emeritus professor of geography at Curtin University in Perth, Western Australia, where he has worked since 1970. He is a historical geographer with research interests in heritage, tourism and rural and regional change.

Alexandre M. A. Diniz is a professor in the graduate program in geography at the Pontifical Catholic University of Minas Gerais, Brazil. His major research interests are in human geography, primarily on agricultural frontiers, crime geography and regional geography with an emphasis on the Brazilian Amazon.

Routledge Research in Historical Geography
Series Editors:
Simon Naylor
University of Glasgow, UK
Laura Cameron
Queen's University, Canada

This series offers a forum for original and innovative research, exploring a wide range of topics encompassed by the sub-discipline of historical geography and cognate fields in the humanities and social sciences. Titles within the series adopt a global geographical scope and historical studies of geographical issues that are grounded in detailed inquiries of primary source materials. The series also supports historiographical and theoretical overviews, and edited collections of essays on historical-geographical themes. This series is aimed at upper-level undergraduates, research students and academics.

Cultural Histories, Memories and Extreme Weather
A Historical Geography Perspective
Edited by Georgina H. Endfield and Lucy Veale

Commemorative Spaces of the First World War
Historical Geographies at the Centenary
Edited by James Wallis and David C. Harvey

Architectures of Hurry—Mobilities, Cities and Modernity
Edited by Phillip Gordon Mackintosh, Richard Dennis, and Deryck W. Holdsworth

Anarchy and Geography
Reclus and Kropotkin in the UK
Federico Ferretti

Twentieth Century Land Settlement Schemes
Edited by Roy Jones and Alexandre M. A. Diniz

For more information about this series, please visit: www.routledge.com/Routledge-Research-in-Historical-Geography/book-series/RRHGS

Twentieth Century Land Settlement Schemes

Edited by Roy Jones
and Alexandre M. A. Diniz

Routledge
Taylor & Francis Group

LONDON AND NEW YORK

First published 2019
by Routledge
2 Park Square, Milton Park, Abingdon, Oxon OX14 4RN

and by Routledge
52 Vanderbilt Avenue, New York, NY 10017

First issued in paperback 2020

*Routledge is an imprint of the Taylor & Francis Group,
an Informa business.*

British Library Cataloguing-in-Publication Data
A catalogue record for this book is available from the British Library.

Library of Congress Cataloging-in-Publication Data
A catalog record has been requested for this book.

ISBN 13: 978-0-36-758526-6 (pbk)
ISBN 13: 978-1-138-05226-0 (hbk)

Typeset in Times NR MT Pro
by Cenveo® Publisher Services

For our wives, Christina Birdsall-Jones and
Ludmila Ribeiro

Contents

List of figures

List of tables

List of contributors

Ana Maria de Souza Mello Bicalho is a rural geographer (PhD, University of London, UK) who is a full professor in the Departamento de Geografia, Universidade Federal do Rio de Janeiro, Brazil, and was chair of the International Geographical Union Commission on the Sustainability of Rural Systems between 2007 and 2012. She has more than four decades of research experience in the Amazon, the Central-West, the Northeast and the Southeast of Brazil and has published articles in *The Journal of Peasant Studies, Geographical Research* and *Horizons in Geography* and has edited a number of books on sustainable rural development.

Mark Brayshay is emeritus professor in the School of Geography, Earth and Environmental Sciences at the University of Plymouth, UK. For more than 40 years, he taught degree-level historical geography. His principal research interest is in the evolving geographies of travel and communication, especially during the early modern period in England and Wales and elsewhere in Europe. He also focuses on the historical geography of Britain's former empire and has engaged in collaborative research on the interlocking directorships of nineteenth and early twentieth century multinational companies. In addition, focusing on the same period, he studies British emigration and overseas settlement.

George Curry (PhD, University of New England, Australia) is professor of geography in the Geography Discipline Group at Curtin University, Western Australia. He has more than 20 years of research experience in Papua New Guinea, where he has worked on a wide range of agricultural development projects with support from the World Bank, the Australia Research Council and a range of government instrumentalities in both Australia and Papua New Guinea.

Alexandre M. A. Diniz (PhD, Arizona State University, USA) is a professor in the Geography Graduate Program at the Catholic University of Minas Gerais, Brazil, which he has coordinated since 2010. He has authored or coauthored numerous refereed publications and was a member of FAPEMIG's scientific chamber (the Minas Gerais State Foundation for

Research Support). His fields of interest are frontier evolution and urban and regional geography.

Tialda Haartsen is a professor of rural geography in the Cultural Geography Department, Faculty of Spatial Sciences, University of Groningen in the Netherlands. Her research on senses of rural belonging and rural change is concentrated around two themes. The first is on the motives and everyday life experiences of newcomers, return migrants, stayers and leavers in rural areas, and how these relate to differences in attachment to, and rootedness in, the countryside. The second is rural transformations and the loss of facilities and services and their impact on quality of life and rural identities. Tialda was born in the Noordoostpolder and is a granddaughter of its pioneers.

Scott William Hoefle (DPhil, University of Oxford, UK) is full professor in the Geography Department of the Federal University of Rio de Janeiro, Brazil, where he lectures in cultural theory and political ecology. He completed his DPhil in social anthropology at the University of Oxford in 1983. Since then, he has undertaken research in North, Northeast, Southeast and Central-West Brazil. Relevant recent publications include "Fishing livelihoods, seashore tourism and industrial development in coastal Rio de Janeiro", *Geographical Research* (2014), "Multifunctionality, juxtaposion and conflict in the Central Amazon", *Journal of Rural Studies* (2016) and (with Ana Maria Bicalho) "Conservation units, environmental services and frontier peasants in the central Amazon", *Research in Economic Anthropology* (2015).

Julia Horsley is a research fellow in geography at the University of Western Australia and has undertaken research on a range of themes related to regional development. Her most recent research is focused on the relationships between government policy and private firms in regional development.

Roy Jones (PhD, University of Manchester, UK) is an emeritus professor in the Geography Discipline Group at Curtin University in Perth, Western Australia, where he has worked since moving to Australia in 1970. He is a historical geographer with particular interest in the areas of heritage and tourism. He has authored or coauthored more than 100 refereed publications, including the Australian chapter in *The Ashgate Research Companion to Heritage and Identity* and the "Heritage and Culture" entry in *The Elsevier Encyclopedia of Human Geography*. He was the human geography editor of *Geographical Research: Journal of the Institute of Australian Geographers* (2001–2009) and is a steering committee member of the International Geographical Union's Commission on the Sustainability of Rural Systems (2012–2020). In 2013, he was awarded a Distinguished Fellowship of the Institute of Australian Geographers.

Gina Koczberski (PhD, University of Sydney, Australia) is a senior research fellow in the Geography Discipline Group at Curtin University in Perth, Western Australia. She has several decades of research experience in Papua New Guinea and the Pacific, where she has worked on projects related to land settlement, transitions from subsistence to commercial agriculture and the role and position of women in development.

Elisangela G. Lacerda (PhD, Pontifical Catholic University of Minas Gerais, Brazil) is an adjunct professor at the Federal University of Roraima, Brazil. Her research interests are the urban and regional geography of Roraima, Brazil.

John C. Lehr (PhD, University of Manitoba, Canada) is a senior scholar in the Department of Geography at the University of Winnipeg, Canada, where he was formerly a professor. He holds an MA from the University of Alberta, Canada, and a PhD from the University of Manitoba, Canada. His research interests focus on the historical geography of Western Canada, particularly on the agricultural settlements of Ukrainians, Jews and Hutterites. His recent publications include *Community and Frontier: A Ukrainian Settlement in the Canadian Parkland* (University of Manitoba Press, 2011) and, with Yossi Katz, *Inside the Ark: The Hutterites in Canada and the United States* (University of Regina Press, 2014).

Steven Nake is an agronomist with the PNG Oil Palm Research Association in Papua New Guinea. He is based in West New Britain Province. He has 19 years of research experience in the field of soil science and plant nutrition and has worked closely with smallholders during this period. In his current capacity as head of smallholder and socioeconomic research, he is involved in both biophysical and socioeconomic research, with the aim of improving smallholder productivity and the social and economic well-being of oil palm smallholder households.

James Richtik is a retired professor of geography at the University of Winnipeg, Canada. His main research area has been agricultural settlement in Western Canada, examining social, governmental and environmental influences on the process. His articles have appeared in *Agricultural History*, *New Zealand Geographer*, *Great Plains Quarterly* and other academic journals.

Michael Roche (PhD, Canterbury University, New Zealand; DLitt, Massey University, New Zealand) is professor of geography in the School of People, Environment and Planning at Massey University, Palmerston North, New Zealand. A graduate in historical geography from the University of Canterbury, New Zealand, he has published widely on New Zealand's forest, agricultural and environmental history, including a series of articles that reexamine the World War I discharged soldier settlement scheme in New Zealand.

John Selwood (PhD, University of Western Australia, Australia) is a senior scholar, and was formerly a professor, in the Department of Geography at the University of Winnipeg, Canada, where he worked for almost four decades. During that time, he researched a wide range of development, settlement and tourism issues in both Western Canada and Western Australia.

Frans Thissen is a human geographer and, since his retirement in 2012, guest researcher at the Department of Geography, Planning and International Development Studies, Universiteit van Amsterdam, Amsterdam, the Netherlands. His research focuses on the quality of life in rural areas in the Netherlands and Flanders, especially the position of young and older people. He was responsible for European Union-financed projects about the life satisfaction of older people and the quality of life in Flemish villages. He is a consultant to local authorities, provinces, community development organisations and research institutes in the Netherlands and Flanders regarding rural community development problems.

Matthew Tonts (PhD, Curtin University, Australia) has longstanding interests in aspects of rural development in Australia. Most of his research has focussed on understanding the interactions of economic development, regional policy and social well-being. Matthew is presently executive dean of the Faculty of Arts, Business, Law and Education at the University of Western Australia.

Acknowledgments

Roy Jones thanks the State Records Office of Western Australia for permission to reproduce the "Western Australia for the Settler" poster.

Alexandre M. A. Diniz acknowledges the support of FAPEMIG's Pesquisador Mineiro Grant – CSA – PPM-00705-16. He also thanks the School of Design and the Built Environment at Curtin University for financial assistance to enable editorial meetings to be held in Perth.

John Lehr thanks the University of Winnipeg for financial assistance for the research on which the article is based.

Mark Brayshay is grateful to Carole Perry of the Northcliffe Pioneer Museum for sharing her knowledge and permitting publication of images from the museum's collection. He also owes thanks to Tim Absalom and Jamie Quinn for drawing the maps and to Mark Birchall for permitting him to consult Sir John Birchall's private archives.

Chapter 7, Reprinted work with permission. Selwood, J and Richtik, J 2003 " 'If they had the will:' Hudson's Bay Sponsored Agricultural Settlers at Vermilion, Alberta." *British Journal of Canadian Studies* Vol. 16, pp. 71–87.

John Selwood and James Richtik thank the Hudson's Bay Company's Archives, Archives of Manitoba, for allowing them to access their collection.

Ana Maria Bicalho and Scott Hoefle acknowledge research funding support from the Conselho Nacional de Desenvolvimento Científico e Tecnológico (CNPq-Brazil), the Fundação Carlos Chagas de Amparo à Pesquisa no Estado do Rio de Janeiro (FAPERJ) and the Institut pour la Recherche de Développement (IRD-France).

Tialda Haartsen and Frans Thissen would like to thank Dr. Henk Pruntel and Dr. André Geurts from Batavialand, Lelystad, for their comments and suggestions regarding the text and figures.

Gina Koczberski, George Curry and Steven Nake acknowledge research funding from the Australian Centre for International Agricultural Research. Special thanks to the smallholders of Hoskins and Bialla, who shared their stories with us, and the OPIC agricultural extension officers, who provided information for the study. The research assistance and support provided by the Oil Palm Research Association is also gratefully acknowledged.

1 Introduction: the lives and legacies of twentieth century land settlement schemes

Roy Jones
Alexandre M. A. Diniz

Introduction

This volume had its genesis in a presentation by one of the editors (Diniz) at a colloquium of the International Geographical Union's Commission on the Sustainability of Rural Systems at the University of Nagoya in 2013. Diniz described the relative failures of the Nova Amazonia land settlement scheme in Roraima Province, Brazil, at the turn of the present century. Notwithstanding the considerable differences in time, space, environment and culture between them, the nature of and the reasons for the failure of the Nova Amazonia project resonated strongly with Jones' knowledge of the much earlier, but equally unsuccessful, Group Settlement scheme in the South West region of Western Australia in the 1920s. These resonances led us to collaborate with three other scholars who were familiar with either the Nova Amazonia (Lacerda) or the Group Settlement (Brayshay and Selwood) schemes to develop a comparative presentation on the two projects for the Commission's 2014 colloquium in Sibiu, Romania, and a subsequent publication on this topic (Jones et al. 2015).

In our publication, we sought to place the fates of these two schemes in both their historical and sustainability contexts, not only to conform to the remit of the Commission but also because the localities where Group Settlement had proved to be such a notable failure (Gabbedy 1988; Brayshay and Selwood 2002) have, decades later, enjoyed considerable success as high-amenity viticultural, touristic and retirement venues (Selwood et al. 1996; Argent et al. 2014). In producing this paper, we had assembled a group of five authors, most of them historical geographers, from four continents (Australasia, Europe, North America and South America), and it was this assembly of collective experience that encouraged us to pursue the issue of the nature and implications of twentieth century land settlement schemes further. More specifically, it was the prospect of, on the one hand, extending this comparative study to further twentieth century land settlement schemes with differing origins, commencement dates and environmental, cultural and political contexts and, on the other, considering the fates, and therefore the sustainability, of a range of such schemes that increasingly engaged our interest. We are grateful to the diverse and dispersed collection of colleagues who have worked with us to create this volume.

The lives of land settlement schemes in the twentieth century

Both agriculture and livestock domestication have taken place over several millennia and in numerous locations across the globe (Ladizinsky 1998). But while, in earlier times, numerous civilisations, from the Mayans to the Dutch, have engaged in major land reclamation and/or development schemes within their own territories, the largest and most long-range land settlement schemes in human history have taken place more recently in the period since the onset of European colonial expansion and the rise of settler societies (Belich 2009). As Belich notes, these movements have given rise to massive expansions in the world's agricultural areas from the prairies to the steppes and the pampas, even though these successes have been accompanied by a number of spectacular and sometimes heroic failures, such as the Darien Scheme in Central America in the seventeenth century (Prebble 2000) and the expansion of farming north of "Goyder's Line" in South Australia in the late nineteenth century (Meinig 1962).

In this volume, we seek to demonstrate that it is possible to consider the land settlement schemes of the twentieth century as a distinct(ive) subset of this long historical process for a range of reasons. In environmental terms, most parts of the earth's land surface that were readily accessible and of high agricultural or pastoral potential had already been settled and developed by 1900. In the course of the twentieth century, transport and communication technologies were being revolutionised in ways that increasingly allowed even the most perishable of agricultural products, from New Zealand lamb to Chilean grapes, to be transported to or, more importantly in the context of this volume, from the ends of the earth. These developments allowed governments and corporations to extend their reach either through imperial connections, as was the case at the beginning of the twentieth century or, certainly by its end, through global supply chains and increasingly globalised business and financial networks.

These developments provided the physical and human contexts within which the land settlement schemes of the twentieth century were played out. Any land that was newly brought into production after 1900 was likely to be far from the markets for its produce and, in many cases, far from the homes of those who first settled it. It was also likely to be environmentally challenging and of relatively limited agricultural potential. Furthermore, these lands were being developed when technological advances were being made rapidly, generating constant changes in the methods of agriculture and animal husbandry; when competition for markets was becoming global; and when political and economic instability was widespread. Those who participated in twentieth century land settlement schemes were therefore vulnerable to a variety of threats including ecological disasters on their environmentally marginal land, market shifts in a liberalising and globalising economy and policy shifts, either by volatile imperial or national governments or by agile corporations seeking to benefit shareholders in the global core rather than agricultural producers at the periphery. For these reasons, we contend that those who involved themselves in twentieth century land settlement schemes faced a mix of challenges and problems that were often different from, but were not necessarily of lesser magnitude, than those faced by the frequently mythologised "pioneers" (Hirst 1978; Stannage 1985; Urgo 1995) of earlier centuries. In discussing these challenges and problems in this volume, in no way do we seek to discount or to minimise the challenges and problems of those, characteristically Indigenous, peoples whose lives were disrupted and whose lands were dispossessed in pursuit of these settlement schemes. Rather, just as volumes such as Laidlaw and Lester (2015) seek to trace the losses, disruptions and survivals of Indigenous populations in the face of settler colonialism, so do we seek to trace the successes, failures and survivals of the settlement schemes and those who participated in them.

The legacies of twentieth century land settlement schemes

As indicated earlier, twentieth century land settlement schemes were characteristically commenced in remote and environmentally challenging locations and were seen as being capable of growing to maturity in extremely dynamic technological, economic, social and political circumstances that, in the event, often rendered the guidelines, or even the rationales, for undertaking such ventures obsolete. For example, in 1905, Western Australia's Royal Commission on Immigration decreed that, "all considerable areas of agricultural land must have a fifteen-mile (24-kilometre) rail service" (Western Australia 1905, p. xxii). This was seen as the maximum distance over which a farmer, using a horse and cart, could be reasonably expected to transport grain or to travel for basic supplies. By the 1920s, the state government had created a relatively dense network of railway lines, townsites and wheat bins across Western Australia's expanding wheat belt (Glynn 1975). By the late twentieth century, a combination of motorised transport, highly mechanised agriculture, rural depopulation and neoliberal economics had rendered most of the rail services and rural service centres, and even many of the wheat bins, surplus to requirements (Jones and Buckley 2017), a process that has been replicated not only in the Australian broad-acre farming regions but also in the Canadian Prairies and many other areas of recent agricultural settlement across the globe.

This example illustrates why a study of the legacies of twentieth century land settlement schemes is a valuable, if not a necessary, corollary of a study of these schemes' lives. Whether or not such schemes were initially or eventually successful, the environments, economies and societies in the areas in which they were established will only be sustainable in the modern world if they can successfully adapt, and continue to adapt, to a wide range of rapidly changing circumstances. The nature, or even the existence, of their legacies will inevitably be diverse. Failure, in some cases, will lead to complete abandonment, as in the case of the farms to the north of Goyder's Line in Meinig's classic study. Initial success or failure may both be followed by the development of completely different agricultural bases from those upon which an agricultural area was first founded. Furthermore, some areas that were initially settled for agricultural purposes may come to depend, to varying extents, on non-farming activities as they progress towards post-productivist futures (Holmes 2002; Almstedt 2013).

Issues of legacy and sustainability will inevitably loom larger in the investigations of those schemes that, today, have a history of several generations of success, failure and/or change. However, many of the case studies included here describe relatively recent, or even current, extensions of the world's agricultural frontier so, before turning to a description of the case studies included in this volume, we include a brief consideration of this complex and, on occasion, contested concept.

Twentieth century land settlement schemes as final frontiers

In our initial submission to the publishers, the proposed title of our work began with the term "Final Frontiers" (albeit followed by a question mark). While we now agree with those reviewers and members of the publishing staff who expressed their reservations about our use of this term, we did not include these words merely as an attention-grabbing device. We sought not only to emphasise the spatial peripherality and the environmental and economic marginality of many of the twentieth century's land settlement schemes, but also to raise the question of whether agriculture and even livestock farming were approaching their spatial limits between 1900 and 2000. What we sought to interrogate was a phenomenon very different from that of the closing of the frontier, as identified by the US census in 1890 (Nash 1980), which perhaps is a convenient date for the purposes of this volume. Rather, we wished, in large part through the case studies assembled here, to consider the extent to which the twentieth century could be seen as a period during which the contribution of areal expansion to the growth of agricultural production came to be of considerably less significance on a global scale, but also as a period when frontier activity continued in a range of dynamic and challenging contexts. For this reason, we preface the regional case studies of land settlement schemes with a chapter reviewing the literature on and theories of frontiers to provide some background information on the processes at work in the regions under consideration in this volume.

Twentieth century land settlement schemes: the case studies

The settlement schemes described in the following case studies cannot present a comprehensive view of the multitude of land development schemes that were instigated over the course of the previous century. The purpose of this collection is rather to depict a range of schemes that differ in age, location, environment, type of protagonist and purpose and to consider them from a range of economic, political and cultural viewpoints, not so much to demonstrate their variety but rather to ask whether, within their diversity, any common elements that have contributed to their initial success, subsequent failure or eventual reorientation can be discerned. In addition, and drawing on the questions raised in our earlier group settlement and Nova Amazonia comparison, we also look for any evidence of lessons learned from earlier land settlement scheme shortcomings that could, or even should, have been taken into account in the planning and implementation of subsequent initiatives. In this section of the introduction, we focus more on the differences among the schemes depicted in these case studies before turning, in our conclusion, to any commonalities that might be discerned among them (Figure 1.1).

The physical environments in which these schemes were undertaken range from equatorial forests to cool temperate grasslands and even include the reclamation of land from the sea. The agricultural products that were the focus of these schemes include palm oil, soybean, wheat, meat, rubber, cheese and butter. The protagonists include colonial, national and

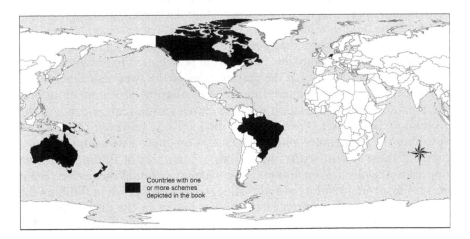

Figure 1.1 Map of land settlement case study countries.

Source: The authors

subnational governments, major corporations and philanthropic bodies, in several cases operating in a variety of what might now be termed public private partnerships. Where the aims of these projects were not purely commercial, they have included various forms of state and national development, the alleviation of unemployment through inter- and intranational migration, the delivery of cultural groups from persecution and the payment of a "debt of honour" to those who had served their country. Given the inevitable intersections between settler societies and relatively recent land settlement schemes, most of the initiatives studied here are from the Americas and Australasia, though these can be contrasted with those from the very different contexts of colonial Papua New Guinea and the Netherlands.

The earliest examples consider schemes that commenced in the latter part of the nineteenth century but extended into the twentieth. The most recent describes a still-expanding active frontier. The following chapters have been sequenced on the basis of the dates at which the various schemes commenced though their lifespans and, still more so, their legacies overlap in time. Overall, these case studies extend over a period from the 1860s to the present.

Matthew Tonts and Julia Horsley address the issue of international knowledge transfer on land settlement schemes. While their main focus is on the Western Australian government's Homestead Act of 1893, which guided much of the state's early twentieth century land development, they consider how the Western Australian legislators and bureaucrats both used and modified the American Homestead Act of 1862 and the Canadian Dominion Lands Act of 1872 to produce land allocation and development frameworks more suited to local conditions (Figure 1.2). Their study also demonstrates how both the traditional "yeoman farmer" ideal and the relatively ready acceptance of the dispossession of Indigenous populations remained in the thoughts and land settlement policies of several governments into the twentieth century.

John Lehr describes how the Jewish Colonization Association provided assistance, albeit in ways that were often inadequate or inappropriate, to settle Jewish families, who were fleeing from persecution in Eastern Europe, on the Canadian Prairies in the first decades of the twentieth century. He documents the organisational difficulties of aligning the aims of an ethnic philanthropic organisation with the bureaucratic requirements of the Dominion Lands Act. Perhaps more distinctively, however, this chapter considers the importance of cultural factors on land settlement. Certainly, many of these Jewish settlers were from urban areas of Europe and were unused to the physical environments in which they found themselves, but it was the insurmountability of the challenges that they faced in fulfilling the social and spiritual requirements of their religion in the dispersed settlements of the prairies that caused many of these settlers to move to Canada's towns and cities.

Michael Roche describes the limited success of soldier settlement on the mountainous and marginal "bush frontier" of central North Island,

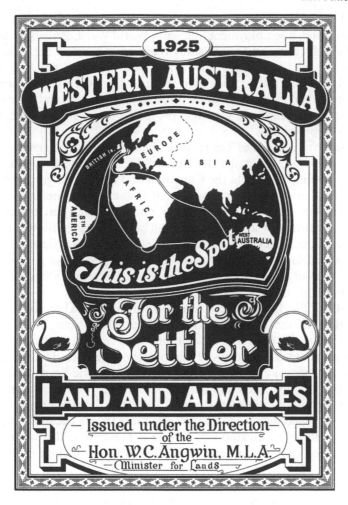

Figure 1.2 "Western Australia for the Settler" poster, 1925.

Source: State Records Office of Western Australia, Cons1496 1925/0573

New Zealand, in the years following World War I. In Australasia, the allocation of land on more than normally favourable financial terms to returned servicemen was seen by governments as one means of repaying a "debt of honour" to those who had served their country. However, expansion of the farming frontier in central North Island was reaching its environmental limits, and the land offered to many of these war veterans was of questionable value. Roche uses both official statistics and individual accounts to depict what he terms as "the end of an era of pioneering closer land settlement".

Mark Brayshay provides a detailed account of the experiences of the "Leeds Group" of 79 British settlers in the south-western forests of Australia in the 1920s and 1930s. The Group Settlement scheme was a cooperative venture between the British and Western Australian governments whereby, at a time of high unemployment following World War I, British settlers were assisted in relocating and establishing themselves as small dairy farmers in what was, at the time, virgin forest land. By focusing on this single group, Brayshay has been able to track their fortunes from their recruitment in the UK through their transport to their forest land allocation and their – largely unsuccessful – attempts to establish profitable farms prior to the onset of the 1930s Depression.

John Selwood and James Richtik describe a complex arrangement between the Hudson's Bay Company, the Cunard Line and the Canadian Pacific Railway, working with both the Canadian and the British governments, by which British migrants were brought out to farm plots prepared for them near Vermilion, Alberta. The parallels with the group settlement scheme described by Brayshay are considerable. The settlers were ill-prepared for a farming existence; they arrived at the onset of the Depression and, on relatively marginal land, they experienced an exceptional run of poor seasons, a concatenation of circumstances that, here as in the Australian South West, led to a high rate of farm abandonment.

Ana Maria Bicalho and Scott Hoefle apply a political ecology approach to the study of the Ford Company's rubber plantations, Fordlandia and Belterra, in Brazil's Para state, which they term "a cemetery for overly optimistic ideas about dominating the rainforest with advanced technical know-how and business savvy". Ford's attempt to parachute an American town and agricultural operation into the Brazilian rainforest was a failure. Blight from the surrounding rainforest readily infested the rubber trees, and the scheme was abandoned in 1945 after little more than a decade of attempted operation. However, Bicalho and Hoefle contextualise this initiative in a wider study of agricultural, and especially latex, production in this region of Brazil.

Tialda Haartsen and Frans Thissen focus on the development of the Noordoostpolder (Northeast Polder), which was drained in 1942, within the framework of the wider scheme by which the Zuiderzee was dammed to create a freshwater lake within which several polders were reclaimed. In a very different physical and spatial context from those of the other case studies in this volume, they describe the physical and social engineering plans and policies of the Dutch government through which it was hoped that a successful agricultural economy and community could be established before tracing the polder's more recent trajectory to a more multifunctional and post-productive present.

Gina Koczberski, George Curry and Steven Nake trace the evolution of a land settlement scheme in West New Britain in Papua New Guinea. In 1967, the Australian colonial authorities acquired sparsely settled customary land and cooperated with a British plantation company to develop smallholder lots to supply oil palm to a processing plant. Settlers from more densely populated parts of the country were encouraged to take up these smallholdings and to engage in commercial, rather than subsistence, cropping activities as part of what was envisaged as a wider shift towards national development and modernisation. Even though this can be considered as one of the more successful land settlement schemes depicted in this volume, this study of half a century of its operation also shows just how slow, economically variable and socially complex these development processes can be in practice.

Alexandre Diniz and Elisangela Lacerda place the development of soybean production in Brazil's poorest and most remote province of Roraima in the contexts of this crop's uptake across Brazil and, more broadly still, of the country's frontier development over the last half century. The Brazilian government has been investing heavily in infrastructure and especially in highway building in Roraima. Commercial soybean production in the province commenced in 1993 and has expanded rapidly in recent years. This is therefore a study of a currently advancing agricultural frontier in a neoliberal and globalising environment.

Roy Jones and Alexandre Diniz then provide an overview of the case study chapters in an attempt to discern the common threads, and even to identify some common lessons, from this wide range of settlement initiatives that were undertaken in diverse environments over a period of massive economic, political, social and technological change.

References

Almstedt, Å 2013, "Post-productivism in rural areas: a contested concept," in L Lundmark and C Sandström (Eds), *Natural Resources and Regional Development Theory*, Institutionen för geografi och ekonomisk historia, Umeå universitet GERUM Kulturgeografisk arbetsrapport, Umeå, pp. 8–22.

Argent, N, Tonts, M, Jones, R, and Holmes, J 2014, "The amenity principle, internal migration and rural development in Australia," *Annals of the Association of American Geographers*, Vol. 104, No. 2, pp. 305–18.

Belich, J 2009, *Replenishing the Earth: The Settler Revolution and the Rise of the Anglo-World 1783-1939*, Oxford University Press, Oxford.

Brayshay, M and Selwood, J 2002, "Dreams, propaganda and harsh realities: landscapes of group settlement in the forest districts of Western Australia in the 1920s," *Landscape Research*, Vol. 27, No.1, pp. 81–101.

Gabbedy, J 1988, *Group Settlement. Part 1. Its Origins, Politics and Administration*, University of Western Australia Press, Nedlands.

Glynn, S 1975, *Government Policy and Agricultural Development: A Study of the Role of Government in the Development of the Western Australian Wheat Belt 1900-1930*, University of Western Australia Press, Nedlands.

Hirst, J 1978, "The pioneer legend," *Historical Studies*, Vol. 18, No. 71, pp. 171–209.

Holmes, J 2002, "Diversity and change in Australia's rangelands: a postproductive transition with a difference?" *Transactions of the Institute of British Geographers*, Vol. 27, No. 3, pp. 362–84.

Jones, R and Buckley, A 2017, "From the horse and cart to the internet: a century of rural connectivity change in Western Australia," *Bulletin de la Societe Geographique de Liege*, Vol. 69, pp. 9–16.

Jones, R, Diniz, A, Selwood, J, Brayshay, M and Lacerda, E 2015, "Rural settlement schemes in the South West of Western Australia and Roraima State, Brazil: unsustainable rural systems?" *Carpathian Journal of Earth and Environmental Sciences*, Vol. 10, No. 3, pp. 125–32.

Ladizinsky, G 1998, *Plant Evolution under Domestication*, Kluwer, The Netherlands.

Laidlaw, Z and Lester, A (Eds) 2015, *Indigenous Communities and Settler Colonialism: Land Holding, Loss and Survival in an Interconnected World*, Palgrave Macmillan, Basingstoke.

Meinig, D 1962, *On the Margins of the Good Earth: The South Australian Wheat Frontier 1869–1884*, Rigby, Adelaide.

Nash, G 1980, "The census of 1890 and the closing of the frontier," *Pacific Northwest Quarterly*, Vol. 71, No. 3, pp. 98–100.

Prebble, J 2000, *Darien: The Scottish Dream of Empire*, Birlinn, Edinburgh.

Selwood, J, Curry, G and Jones, R 1996, "From the turnaround to the backlash: tourism and rural change in the Shire of Denmark, Western Australia," *Urban Policy and Research*, Vol.14, No. 3, pp. 215–25.

Stannage, T 1985, *Western Australia's Heritage: The Pioneer Myth*, Monograph Series No.1, University Extension, University of Western Australia, Nedlands.

Urgo, J 1995, *Willa Cather and the Myth of American Migration*, University of Illinois Press, Urbana and Chicago.

Western Australia, Parliament 1905, "Report of the Royal Commission on Immigration and Land Settlement" in *Minutes and Votes and Proceedings of the Parliament*, Western Australian Government Printer, Perth.

2 Conceptualizing the frontier

Alexandre M. A. Diniz

Introduction

The frontier has an enduring place in the popular imagination with various literary and cinematographic accounts portraying the mystique of "civilization's" encounters on the fringe with indigenous peoples, pristine and abundant lands and adventurous characters. Frontier expansion has occurred throughout human history and accounts for the human occupation of nearly all habitable parts of the planet. The forms of colonization have varied greatly, ranging from the territorial extension of hunter-gatherer groups to the conquest and exploitation of entire continents by complex, industrial states (Lewis 1984). Today, the frontier lives on and its continuing appeal stems from the ongoing movement of populations to less inhabited realms, which continues to have an impact on socioeconomic and environmental systems and, for good or ill, to be an important development force (Brown and Sierra 1994; Barbier 2004, 2005). Before we turn to the specific studies in this volume, which offer various accounts of settlement schemes leading to frontier movements in the twentieth century worldwide, it is necessary to build a conceptual framework against which to compare and contrast the experiences reported therein. We do so by engaging in an attempt to define the frontier and to summarize some of its various theoretical interpretations.

The frontier

From the outset, it is important to recognize that the term "frontier" represents differing realities and evokes conflicting meanings. It is both a concept and a process, and it is operationalized in a multitude of situations and on a wide range of geographical scales including:

> ... the peripheries of settlement as well as a densely settled area of mixing and exchange, a barrier and a zone of transition, an open zone of expansion as well as a closed fortified defence line, the ordinary edge of administrative units as well as the limits of a civilization, the edge

of an empire and the edge of a land-holding, the extensive expanses of steppes dividing various nomadic groups and the narrow boundary between states, the marches of the Empire and an overseas colony.

(Janeczek 2011, p. 7)

Although its literal meaning implies the front tier of a territory—that part of a region that fronts another region (Green and Perlman 1985, p. 4), the term frontier is often used metaphorically (e.g., frontiers of science, frontiers of knowledge, frontiers of the soul, etc.), adding further levels of complexity to its meaning and taking it beyond the spatial aspects of human organization that are the primary focus of this book.

Another source of confusion lies in the fact that "frontier" is often mistaken for and confused with the concept of a "boundary". But, while boundaries are conceived as lines, frontiers are characterized as zones. Boundaries represent demarcation lines, indicating the well-established and agreed-upon limits or bounds of a given spatial unit. They are intrinsically inner-oriented and centripetal and act as a separating device. Conversely, the frontier has an outer-oriented centrifugal meaning. It is a borderland into which people move, and it is therefore an integrating element between land already possessed and that which is being claimed or aspired to (Feuer 2016).

Green and Perlman (1985) further elucidate the specificities of these two concepts by emphasizing their complementary nature. Accordingly, frontier studies focus on the peripheries or edges of particular territories and the characteristics of those groups occupying these spaces, seeking to understand the motivations behind their demographic, political and economic expansions into new habitats and their corresponding impacts on indigenous societies and ecological systems. Boundary studies, on the other hand, explore the interplays occurring at these territorial edges, focusing on the social, political and economic factors guiding the transactions between those societies on each side of the boundary through, for example, trade and immigration.

Feuer (2016) has summarized the contribution of several authors on the characteristics of frontiers, portraying them as dynamic social networks, covering large geographical areas, where culturally diverse societies are in contact. Frontiers encompass three distinct and complementary aspects: a geographical element, extending through a physical area; a cultural aspect, in which societies that once were separated and distinct come into contact; and a relational dimension, by which the exchanges among such societies may commence, develop and eventually mature. As a temporal phenomenon, the frontier arises with the first arrival of permanent settlement and ceases to exist only when the limits of growth are achieved and when a single authority has established political and economic hegemony over the area. Along these lines, Mayhew (1997, p.184) furthers the concept by postulating that "the frontier represents the part of the country which lies on the limit of the settled area". Therefore, at any given time, a frontier settlement marks the furthest advance of a cultural group into new territory.

Martins (1996) envisions the frontier as an area marked by three well-defined domains: the demographic frontier, the economic frontier and the expansion front. The demographic frontier represents the limit of the settled area beyond which lie indigenous populations and pristine areas. The economic frontier is marked by the conversion of land into merchandise and the transformation of its former, characteristically subsistence-based, inhabitants into waged labourers. Between the demographic and economic frontiers lies the expansion front, a transition zone marked by populations yet to be incorporated by the economic frontier. The expansion front represents the movement of "civilized" populations and economic activities. These encompass not only commercial agriculturalists and other entrepreneurs and government agents, seeking to establish cities and political and judicial institutions, but also poor, non-indigenous populations operating at the fringes of the market economy. The expansion front is characterized by the spread of commerce and exchange networks in which money is frequently absent. Markets are often based on monopolistic village entrepreneurs who control not only commerce, but also worker relations through peonage and enslavement by debt.

These multiple views contain certain common aspects. Frontiers are created in regions where a given social group expands its domain or establishes colonies, characteristically to the detriment of any indigenous groups. It involves the migration of people into new lands, bringing these areas and their inhabitants within the social and economic domain of the expanding society. Furthermore, during the period when such lands are active frontiers, they often tend to lie beyond, or at least at the extreme limits, of state control; thus, both these territories and the people inhabiting them lack effective formal or even sociocultural regulation. Finally, frontiers are inherently perceived as transitory spaces undergoing structural transformations and moving through an evolutionary phase before achieving a new equilibrium.

The academic study of frontiers commenced with the work of Frederick Jackson Turner (1893), who explored the impact of the frontier on American culture and society. Turner's seminal work, however, was criticized for its limited application in other frontier situations, inspiring a plethora of definitions and models of frontier evolution (Feuer 2016), some of which are reviewed here in order to provide a theoretical framework against which the case studies presented in this volume can be viewed.

Theoretical formulations

The contribution of Frederick Jackson Turner's seminal work, "The significance of the frontier in American history", presented at the World's Columbian Exhibition in Chicago, is undeniable. Turner related phases of cultural and economic development to a sequence of changes in geographical landscapes to produce an evolutionary description of American frontier settlements in which change could be seen as a result of ongoing adaptive responses to the frontier environment. The end result of this process of adaptation was the creation of societies that were fundamentally different from those of the settlers' homelands (Lewis 1984). Turner incorporated the concepts of invasion and succession into his interpretation. Accordingly, successive waves of settlement (or *invasion*) each pushed the frontier farther west and created a distinguishable buoyant, self-reliant and enterprising American character. Each incoming group built upon the experience of their predecessors, with each stage ultimately building towards a mature settlement system (or *succession*) (Warf 2006).

Subsequently scholars have expanded Turner's distinctly American view, considering the frontier in cross-cultural perspectives and giving rise to a series of typologies and evolutionary models, all attempting to describe and even to explain the evolution of frontiers. The exploration of frontier areas worldwide has been an important component of the evolution of many societies, and this process has therefore inspired the construction of various theoretical frameworks. Nevertheless, all of these tend to envision frontier areas as highly volatile environments marked by intense structural changes. Frontier theoretical formulations can be broadly classified with respect to scale (Brown and Sierra 1994), and the discussion of frontier models provided here considers these in terms of their global, regional and place-specific natures.

Global perspectives on frontiers largely emanate from scholars working with the World-System theory of development. They elaborate on the transformations undergone by social groups while moving from precapitalist to capitalist-oriented forms of production and on the articulations between core and peripheral areas. Regional frameworks focus on spatial diffusion and on the changing hierarchical functions of sets of places across a given area, whereas place-specific approaches are focused on the evolution of particular areas of settlement over time. We now turn to a scale-based discussion of some of these major frameworks.

Global perspectives

The notion of "development" is somehow inherent in all theoretical formulations within the frontier literature, especially among those working at the global scale. However, the literature on development is extensive, fragmented and contradictory, and no agreement exists therein around the notion of "development". For some, development is regarded as economic and material progress, defined as the capacity of an economy to grow consistently for a number of years (Cotta 1978). Todaro (1989, p. 62) introduces a social component by stating that development is a "multidimensional process involving the reorganization and reorientation of entire economic and social systems". Wilber and Jameson (1992, p.14) present an interesting but vague notion of development as the "progressive emancipation of peoples and nations from the control of nature and from the control of other peoples and nations". Johnston et al. (1994, p.128) go even further, introducing a freedom-like dimension by stating that development is the process that "enables individuals to make their own histories and geographies under the conditions of their own choosing".

The social science literature includes a number of frameworks that identify the preconditions for development and describe the evolution of this process. These views can be broadly organized into four major paradigmatic groups: stage formulations, dual society models, historic-structural perspectives and the neoliberal approach. The World-System perspective is particularly relevant to all of these major development frameworks.

A group of studies have explored frontiers, not as isolated entities but as integral components, of a wider economic system, entailing long distance, if not global, spatial processes. This school of thought falls under the historic-structural development perspective, and many of its ideas stem from Immanuel Wallerstein's text, *The Modern World-System* (1974, 1980). Accordingly, the world economy is seen as a dichotomous entity based on the international division of labour, with the *core European states* at its centre and *peripheral areas* at its boundaries. Exchange between these peripheral areas and the core states is characterized by a "vertical specialization" involving the movement of raw materials from the former to the latter and the movement of manufactures and services in the opposite direction (Gould 1972, pp. 235–6).

One salient aspect for the study of frontier areas that has been brought to light by the World-System model is the fact that, as the world economy is continually expanding, its geographical structure periodically changes to accommodate new growth, enlarging its boundaries and thereby creating and, in time, incorporating new frontier areas. Within this perspective, peripheral areas in newly colonized regions tend to occupy the outer limits of settlement. Their economic production tended to be strictly controlled by the core states, being restricted to lower-ranking goods that were integral to daily life in the economy's core but that were produced by less well-rewarded labour on its periphery (Wallerstein 1974, 1980, p. 302). Such goods usually consisted of raw materials that could be exchanged for manufactures and services from the core states (Gould 1972, pp. 235–6).

Within this framework, factor migration is an essential part of the asymmetric relations established between core and periphery and is necessary to provide effective control of the supply of colonial resources and to increase the potential for trade. Nonetheless, Gray (1976, pp. 127–9) argued that, although factor migration is a characteristic of colonization, the nature of factor settlement varies between tropical and temperate areas.

The colonization of tropical areas involves the introduction of technology and capital in order to produce commercial crops in a more efficient way or to carry out various extractive activities. Foreign residence tends to be temporary and strongly reliant on imports to reproduce many of the conditions of home life and to establish savings in homeland banks, resulting in significant transfer flows from the colony to the mother country. Conversely, in temperate climates, many important exports are often competitive with domestic production in the core state. Production therefore requires migration of labour from the homeland that is frequently encouraged by differences in wages and labour markets between the two regions. This more permanent resettlement in temperate colonies reduces the leakages of funds for imports to the mother country and increases the reinvestment of these funds in real assets in the colony. Nevertheless, if sufficient comparative advantage is achieved, both trends encourage economic development (Gray 1976, pp. 129–30).

Building on these ideas, Steffen (1980) devised a twofold typology of frontiers. Cosmopolitan frontiers are highly dependent upon the parent state, which manipulates all colonials' activities, suffocating attempts at self-rule and allowing virtually no opportunities for indigenous development. The economy of cosmopolitan frontiers is highly specialized and their goods are noncompetitive in global markets. On the other hand, insular frontiers enjoy a greater degree of freedom, and indigenous development is more likely to occur. Their economies are far more diverse and long term in nature, involving specialized and generalized agricultural products, and they are more competitive in nature. It is important to stress that the success of such frontiers is associated with a greater adaptation of settler groups to local conditions. Steffen (1980) emphasizes that these forms of colonization are not necessarily mutually exclusive developments, sometimes occurring simultaneously in the same area. Another aspect stressed by Steffen (1980) is that these types of colonization generally take place in sequence with the cosmopolitan frontier being generally associated with initial stages of colonization.

Drawing on the North American experience, Vance (1970) proposed the mercantile model, a framework of settlement and growth of new lands whereby the source of change is external to the developing area. According to Vance, the initial occupation of North America took place through the development of selected settlements, which were strongly connected to Europe. The dynamics of growth were exogenous and based on trade. As the scale of the trading system enlarged, mercantile towns grew, which, in turn, increased the demand for hinterland provisions from those towns. This greater demand led to the growth of smaller settlements located in the hinterland, in turn promoting their growth and, through this, the evolution of the entire urban system. Because the mercantile model is based on the wholesaling function, it became progressively less relevant after 1850. As a result of American industrialization, many functions that had traditionally been seen as components of wholesaling, such as transportation, finance and insurance, became organizationally and spatially specialized while much of the wholesaling activity became internalized in manufacturing firms (Meyer 1980).

Lewis' (1984) contribution is also noteworthy because it focuses on those insular frontier systems that were associated with the colonization of new areas by European powers. In this context, changes in the size, composition, direction and nature of settlements could only be fully grasped at the regional scale, given the emphasis on their articulations with the mother countries. Inspired by many frontier evolution models, Lewis (1984, pp. 25–7) proposed a framework that emphasizes complex processual changes resulting from the expansion of an intrusive society into a new territory over a period of time. These occur over six phases.

During the *establishment* phase, the colony is created as a permanent settlement sustained by the production and export of competitive goods primarily for markets in the parent state. As accessibility is crucial to successful commercial agricultural production, settlement will follow the transportation networks, characterizing the *transport and spatial patterning* phase. The frontier tends to expand in response to increasing external demands for goods and improvements in the methods and organization of the transportation system, comprising the *expansion* phase. With time and the intensification of relations with the parent state, the settlement pattern of colonial areas changes in response to increasing population density and economic complexity reaching the *settlement pattern* phase. During the *organization of activities* phase, low population density leads to a dispersed settlement pattern, mirroring the organization of activities in the colony. The smaller number of settlements results in an abbreviated hierarchy of places that tends to concentrate social, economic, political and religious activities at focal points, called frontier towns. The settlement pattern and social relations of production may vary with the staple crop grown. Lastly, during the *colonization gradient,* the hierarchy of settlement displays a pattern of increasing socioeconomic complexity. As the frontier region expands, the spatial patterning of the colonization gradient is likely to be repeated in newly settled areas.

Regional approaches

Regional development theorizations tend to be focused on the spatial diffusion and urban hierarchical functions of a set of places across a given area. Leyburn (1935) examined the spatial and processual elements of a number of frontiers around the world. These were understood as regions located on the outer fringe of societies where pioneer groups made adaptive changes in order to survive in pristine environments. Their different forms of response-adaptation were classified as four subtypes based on the motives of the pioneers and the nature of the various colonies, namely: small farm, settlement plantation, exploitative plantation and camp frontiers. While the first two are characterized by permanent settlements, the remainder involve only transient male settlers. Leyburn's (1935) model recognized important patterns and differences among settlements, thereby establishing a nomothetic view of frontiers.

Building on the Brazilian Amazon experience, Neiva (1949) envisions frontier areas as being possessed of demographic and economic fronts. The process of occupation of what were perceived of as pristine areas was marked by the arrival of the "demographic frontier" composed of small-scale producers, such as farmers and artisans, who tended to be the pioneer agents in frontier areas, before the arrival of the "economic front", marked by enterprises and large-scale market-oriented producers, made its presence felt. During the earlier stages of this evolutionary process, land was relatively abundant and the demographic front thrived. With the arrival of the economic front, conflicts took place over local resources due to the tendency of capitalist enterprises to expand and control the means of production. According to this model, these two fronts inevitably clashed at some point in time, engendering the out-migration of small producers. Here, migration is treated as a mechanical and ready response. Neiva's notion that places within the frontier are transformed from "demographic" into "economic" frontier areas is intriguing and has inspired a series of frontier development proposals and migration studies in the Amazon region.

By the 1950s, geographers had begun to treat frontiers as zones adjacent to formal boundaries, beyond which lay the unsettled portions of territory deemed to be under the effective control of a given state. Frontiers represented zones of transition "stretching from the edge of the state core to the limits of its expansion" (Kristof 1959, p. 274; Weigert et al. 1957, p.115). Building on this perspective, Prescott (1965) made a distinction between settlement frontiers – regions of colonial expansion, and political frontiers – claimed borderlands between two states. Prescott (1965, p. 55) also identified two types of settlement frontiers: "primary frontiers, representing settlement regions at the de facto limit of a state's authority; and secondary frontiers, designating those areas originally passed over during initial expansion and settled only later when less suitable land became desirable due to population pressure".

Based on the frontier expansion of northern Sweden during the late eighteenth and early nineteenth centuries, Bylund (1960) formulated a series of deterministic models in which the spread of settlement over a theoretical landscape was seen to be governed by a set of attraction values. According to his framework, the attraction exerted by an area was inversely proportional to its distance from a road, church and marketplace. Families were the primary agents in the colonization process, and new areas were occupied by the children of the first pioneers, generation after generation.

Hudson (1969) postulated that human settlements in frontier areas could be treated as analogous to the spread of plants and animals. According to this perspective, the diffusion of rural settlement could be divided into three phases: colonization, spread and competition. In biology, the first of these phases is characterized by species invading a new area, while its counterpart in settlement is the long-distance immigration of colonists from outside the region. The second phase is characterized in biology by regeneration and short-distance dispersal through root development or the hatching of larvae from clutches of eggs and in settlement by secondary colonization of the clone type. Finally, in the competition phase in biology, further diffusion is checked by the physical environment so that weaker individuals are killed off by their stronger neighbours. The process of competition is also seen in the struggle among settlements to control and, if possible, increase their hinterlands. Towns compete with one another for customers, extending their trading areas as far as possible. Through the process, larger settlements and trade centers absorb smaller ones, thereby lowering the density of settlements.

Based on their studies of Ecuador, Casagrande et al. (1964) identified certain regularities associated with local agricultural colonies along the lines of a "colonization gradient". This term describes a process by which the cultural system of the homeland is extended and replicated on the frontier whereby it is integrated with the homeland. In this framework, colonization is conceptualized, again by analogy, with the biological process by which an organism establishes itself in a new ecological niche. The niche wherein the colony expands depends upon the exploitation of local resources and exhibits various forms of adaptation. Thus, the colonization process leads to an eventual simplification of the social, economic and political systems of an initially intrusive society. The degree to which cultural impoverishment ensues in individual cases varies according to distance to the frontier edge and extends through a gradient of settlement types. These types exhibit varying sociocultural complexities that are also characteristic of the evolutionary process of frontier settlements through time.

Muller (1977) describes the staged evolution of the nineteenth century North American frontier. During the *pioneer periphery phase*, the frontier has a series of local service centres. Some of them become more important than others, performing different urban functions largely due to their location at nodal points of the incipient transportation system. During the second stage, the *specialized periphery*, towns become differentiated in performing a broad range of economic functions due to variations in their resource endowments, comparative advantages and inertia. Sharp distinctions in their size and importance also arise at this phase. The third stage is that of the *transitional periphery* during which selected towns become closely connected with the national urban system and emerge as dominant regional centres. The growth of industrial activities and unique functional specializations are seen as important elements in this stage of the transition.

Also based on the North American experience, Meyer (1980) focuses on the mechanisms underlying differential growth among urban centres. A main element is the control over the exchange of goods and services, which is associated with economic specialization, transportation, communications and agglomeration economies. The movements of people and commodities are also elements of this scheme. In an early stage of frontier urbanization, specialization is minimal; entrepreneurs control exchange in a local hinterland only, and physical movements are limited. Over time, selected towns acquire specialized economic activities, but entrepreneurs serving local hinterlands still function, which gives rise to an urban hierarchy on the frontier. Finally, as the frontier becomes settled, its main towns become gateways, chief suppliers of goods and services to the next frontier zone.

Lonsdale and Holmes (1981) examined the settlement systems of sparsely populated regions in the US and Australia that, in many ways, still exhibit the conditions of frontier areas, and came up with a list of similar characteristics: they are isolated and exist in inhospitable and unattractive physical environments; their economic activities are restricted to the exploitation of those natural resources yielding the highest returns; basic services are either unavailable or too costly; highly mobile, gender and age unbalanced populations exhibit low dependency ratios; indigenous groups survive in scattered areas and face cultural identity loss and difficulties in assimilating with the encroaching society; and, although technology may minimize isolation, it cannot substitute for the complex daily social interactions enjoyed by the inhabitants of more densely populated areas.

Brunn and Zeigler (1981) analyse the forces behind the initiation of settlement in previously uninhabited territories and the further settlement of sparsely populated regions, stating that these forces vary from area to area. Nonetheless, they were able to identify a certain number of elements that control these human movements, ranging from structural to individual characteristics. In terms of economics, the factors influencing frontier movements are seen as the difficulties in making a living and providing for basic human needs (food, fuel, construction materials). The physical aspects entail restrictions imposed by the physical environment, such as lack of water resources, unsuitable soils and extreme climatic conditions. Institutionally, frontier movements are constrained by public institutions or policies that prohibit or limit settlement or development. Finally, personal issues controlling frontier movements are associated with fears of isolation, of empty spaces, of frontier cultures, of aboriginal groups, in short a fear of the "new" for those residing in compact urban areas.

Place-specific approaches

The articulation of modes of production is a body of thought in development studies that belongs to the historic-structural school and represents a criticism of earlier forms of dependency theory. Here, the focus of explanations of underdevelopment is shifted away from external relations and markets and towards the concept of local modes of production in an attempt to construct an alternative understanding of the development process (Ruccio and Simon 1988). The articulation of the modes of production approach is briefly reviewed here because it provides a conceptual framework for the analysis of the transition between noncapitalist and capitalist modes of production, which is so vividly present in most place-specific frontier evolution models.

According to Hindess and Hirst (1975, p. 9), a mode of production can be conceptualized as "an articulated combination of relations and forces of production". The relations of production define the specific way in which surplus labour will be appropriated within any given society, as well as the way in which the means of production will be operationalized. Surplus labour is the result of production that exceeds the levels of the simple reproductive requirements of labour and thereby produces surplus value (Hindess and Hirst 1975). Forces of production, on the other hand, are defined by the modes of appropriation of nature or, more simply, the labour process through which a given raw material is transformed into a finished product.

Taylor (1979) argued that the reality of most developing areas is dominated by an articulation of at least two modes of production (capitalist and noncapitalist), in which the capitalist mode is becoming increasingly dominant. Accordingly, the evolution of capitalism encompasses the creation, maintenance and the breakdown of noncapitalist modes of production (Ruccio and Simon 1988). Along these lines, Rey (1975) provides not only the earliest, but also the more elaborate, proposal. Rey (1975) conceives of the articulation of modes of production as proceeding in three stages. First, the capitalist mode of production is imported into noncapitalist peripheral societies and proceeds to reinforce and even to create noncapitalist modes of production. During this stage, capitalism obtains raw materials and labour from noncapitalist social formations, while slowly introducing cash transactions and industrialized goods. Second, capitalism takes root and subordinates the local noncapitalist modes, while still making use of them. This stage is marked by the expansion of capitalist relations. In this process, peasant agriculture and handicraft industries are progressively eliminated, while a wage labour force is created. The break with the land, however, is initially partial and often seasonal because capital tends to penetrate agriculture on a gradual basis. Thus, capitalism begins to ensure its own labour supply. Finally, at a certain point in the evolution of these economies, the capitalist mode of production supplants all noncapitalist forms, a stage yet to be reached in many, if not most, developing countries.

A typical articulation of mode of the production school work is Goodman's (1977) study on the expansion of frontier areas in Northeast Brazil in the 1950s and 1960s. Goodman (1977) attributes the regional backwardness of Northeast Brazil to the spatially unbalanced nature of capitalist growth, which results from the exploitative articulation by the central of the peripheral areas of Brazil. One of the key themes in this articulation is frontier expansion and human mobility. In this process, thousands of individuals, in escaping the economic hardships of Central Brazil, spontaneously expanded the limits of the agricultural frontier in Northeast Brazil.

Land was initially cultivated by families on a subsistence basis, introducing the peasant mode of production to the region. By clearing and cultivating land, peasants increased its economic value and facilitated the introduction of permanent cash crops or conversion to pasture. This work was instrumental for the expansion of capitalist ventures in the region. As time progressed, greater commercial activity and the closer integration of regional markets stimulated agricultural specialization and the penetration of the capitalist mode of production on the frontier. Eventually, subsistence holdings were consolidated into large farms and ranches, and the size of the resident labour force adjusted accordingly.

Goodman's (1977) analysis contends that the establishment and maintenance of noncapitalist modes of production in the periphery is a corollary of the process of capitalist expansion at the centre. The development of subsistence agriculture in Northeast Brazil not only promoted the taming of the land that could then be taken over by large agribusinesses, but also accentuated economic inequalities in Central Brazil, which had the potential to promote social instability. Labour migration is an important component of this formulation because migration provides a means of expanding the frontier, and the transformations introduced by migration catalyse capitalist expansion. Thus, Goodman's (1977) central thesis is that agricultural expansion involves the initial adoption of extensive methods that demand elastic supplies of land and labour, the ownership of which is subsequently consolidated by the formation of large estates. Within this context, the development of land at the frontier not only creates the preconditions for the production of these large land holdings but also permits the survival of noncapitalist modes of production as a necessary precursor of capitalist expansion.

Place-specific frameworks are focused on the evolution of particular settlements over time. Theorists working at this scale postulate that frontier settlements evolve through a set of hierarchical stages, moving from a precapitalistic type economy, characterized by the lack of labour and land markets, to more capitalist-oriented forms of production. These models also tend to emphasize the clashes among different interest groups over frontier resources, which lead to waning in-migration and the displacement of earlier settlers by incoming entrepreneurs.

Henkel (1982) and Findley (1988) propose similar frontier settlement models that likewise emphasize the behaviour of individuals. They postulate that the evolution of frontier settlements occurs in three major stages: pioneering, commercialization and abandonment/consolidation. During the *pioneering stage*, settlers are concerned primarily with occupying the land and bringing it into production. Colonists rely on their large families and on each other to clear the land. In order to fulfil their cash requirements, colonists engage in wage employment at nearby farms. The *commercialization stage* is marked by the construction and improvement of transportation linkages, which provide access to markets and stimulate production. The *abandonment/consolidation stage* is marked by a bifurcation of land tenure. In the process, earlier arrivals sell their land, motivated by debts, lack of capital or environmental constraints, while the better-off consolidators buy the improved land, adding to their existing holdings. As land holdings devoted to livestock or commercial crops require fewer year-round labour inputs, the population retention potential of the colonization zones becomes compromised. As colonists who remained in the area still depend on waged labour for their survival, they tend to leave in search of another plot of land. Thus, land concentration leads to extensive population turnover and further migration, not only among those who sold their plots but also among those who are willing to stay.

Foweraker's (1981) framework emphasizes the process by which frontier areas become progressively connected to national economies. The transformation of pristine frontier areas into "productive societies" is understood as a three-phase transition with noncapitalist, precapitalist and capitalist society stages. This idea of transition implies changes in production and market relations with regard to goods, land and labour (Foweraker 1981, p. 27). At the *noncapitalist stage*, the frontier economy, based on resource extraction, is remarkably isolated. The sphere of exchange is limited to outside markets for one or two locally produced commodities. This stage is marked by the lack of markets for land or labour and social relations of production are mainly servile. There is an emerging commodity sector, which will eventually favour the occupation of the region by peasants.

The expansion of the commodity sector generates greater migratory flows into the region and more intensive extractive activity, which characterizes the second *precapitalistic stage* of frontier expansion. Land begins to be bought and sold, but prices represent only what is on the land, rather than the land itself. These changes bring about more stable production of commodities and an emerging market for land; nevertheless, labour markets are still absent. Social relations of production are established by the growing commodity sector or are mixed forms of servile and capitalist relations.

The *capitalist stage* of frontier expansion is characterized by intense migratory flows into the region and established access to the national economy. The economy is no longer based on extraction but rather on agriculture, which is becoming increasingly capitalized. Land prices rise and land ownership becomes concentrated. Capitalist relations of production are dominant and a labour market is finally achieved. As the frontier moves into its final stage of evolution, economic activity generally becomes more differentiated, resulting in a more complex social division of labour (Foweraker 1981, pp. 27–39).

Based upon the Brazilian experience, Browder and Godfrey (1990) developed the "Amazonian landscape change and urban transition" model, postulating that settlement typically occurs in a progressive sequence, promoted by a gradual incorporation of a dependent peripheral region into the larger national economy, whereby isolated rural settlements progressively become urbanized areas connected to regional urban networks. The evolution of the Amazonian frontier can be divided into five stages: native subsistence economy, resource-extractive frontier, pioneer agricultural frontier, relict frontier and finally a phase of urban primacy and rural depopulation.

Based on a comparison of Latin American tropical forest settlement experiences, Marquette (2006) considers the prospects for frontier development and sustainability. According to Marquette (2006), several studies from the 1980s share common patterns and document phases of adaptation to the frontier that many settlers and frontier regions may experience over time. The first phase lasts for up to five years, during which time settlers attempt to gain a foothold on the frontier by engaging in subsistence agriculture. During this adaptation phase, settlers endure many hardships, resulting in low levels of welfare. The second phase extends from around the fifth to the tenth year of settlement. Having established themselves on the frontier, settlers may start to experiment with new activities beyond subsistence agriculture in a search for economic diversification and betterment. During this experimentation phase, the aspiration is that advances will outweigh setbacks and that welfare will therefore increase. After ten years, settlers enter the "consolidation phase" in which diversification is intensified and resources are used more efficiently, leading to higher levels of economic welfare. This economic differentiation and development bring about structural changes on the frontier. These relate to the region's overall socioeconomic, political and organizational conditions—though, for the individuals involved, these are viewed through the lenses of: farm characteristics, "history" or significant events and changing household socioeconomic and demographic characteristics.

Conclusion

Even this brief review of the literature demonstrates that the frontier is a highly polysemic term, encompassing a plethora of meanings. As a spatial process, it represents the movements of intrusive social groups into territories that are unsettled or are characteristically occupied by small groups with limited powers of resistance. As a historical process, frontier expansion involves the displacement and dispossession of hunter-gatherer groups and subsistence agriculturalists by increasingly commercialised farmers and other settlers. As a cultural process, the frontier juxtaposes invasive social groups, either with the wilderness—resulting in the adoption of a series of adaptive responses to unfamiliar environments, or with autochthonous populations—leading to cultural clashes, alterity and practices of alienation, integration or assimilation. As an economic process, frontier expansion represents the diffusion of mercantilist/capitalist production systems into new areas, engendering core-periphery type interactions at various scales: between more advanced and less developed parts of countries, between imperial states and their colonies or between cores and peripheries on a global scale. As a social process, frontiers can operate as safety valves, relieving population pressures on limited resources in traditionally settled areas while creating social mobility opportunities for newcomers. As an environmental process, frontier expansion has led to profound transformations of natural systems and the conversion of pristine ecosystems into resource-based units. As a geopolitical process, frontier expansion has promoted the de facto occupation of contested areas, thereby providing a rationale for the extension of political sovereignty and hegemonic power over vast regions.

Given its multitude of causes, impacts, motivations and meanings, frontier expansion has inspired a series of theoretical interpretations, covering a vast array of factors working at multiple levels of analysis and ranging from local to global perspectives. Overall, the theoretical formulations working at the settlement and regional scales envision the frontier as a highly volatile environment marked by intense structural changes. Nevertheless, while recognizing that many frontier settlements are transient in nature and disappear in time, the majority of the conceptual interpretations postulate that frontiers tend to undergo an evolutionary process, passing through a series of stages, moving from a few isolated settlements to areas marked by well-defined hierarchies of places. Accordingly, a frontier opens with the first contact between the intrusive group and the new area; and it closes when a single authority has established political and economic hegemony over the zone and when the frontier has been fully integrated into the economic and political systems of a hegemonic power group. Another common trait of the closing of frontiers is an increase in the restrictions imposed on access to local resources, which either become completely exploited or dominated by first-wave immigrants.

It is in the context of these overall ideas that the case studies in the following chapters can be viewed. A number of these ideas will be revisited in the final chapter in a wider assessment of the fates of several land settlement schemes over the course of the twentieth century.

References

Barbier, E 2004, "Agricultural expansion, resource booms and growth in Latin America: implications for long-run economic development", *World Development*, Vol. 32, No.1, pp. 137–57.

Barbier, E 2005, "Frontier expansion and economic development", *Contemporary Economic Policy*, Vol. 323, No. 2, pp. 286–303.

Browder, J and Godfrey, B 1990 "Frontier urbanization in the Brazilian Amazon: a theoretical framework for urban transition", *Conference of Latin American Geographers*, Vol. 16, pp. 56–66.

Brown, L and Sierra, R 1994, "Frontier migration as a multi-stage phenomenon reflecting the interplay of macroforces and local conditions: the Ecuador Amazon", *Regional Science*, Vol. 73, No. 3, pp. 267–88.

Brunn, S and Zeigler, D 1981, "Human settlements in sparsely populated areas: a conceptual overview, with special reference to the U.S." in R Lonsdale and J Holmes (Eds), *Settlement Systems in Sparsely Populated Regions: The United States and Australia*, Pergamon, New York, pp. 14–52.

Bylund, E 1960, "Theoretical considerations regarding the distribution of settlement in inner north Sweden", *Geografiska Annaler* B, Vol. 42, pp. 225–31.

Casagrande, J, Stephen, I and Philip, D 1964, "Colonization as a research frontier" in A Robert (Ed), *Manners Process and Pattern in Culture, Essays in Honor of Julian H. Steward*, Aldine, Chicago, pp. 281–325.

Cotta, A 1978, *Dicionário de Economia*. Publicações Dom Quixote, Lisboa.

Feuer, B 2016, *Boundaries, Borders and Frontiers in Archaeology: A Study of Spatial Relationships*, McFarland, Jefferson, NC.

Findley, S 1988, "Colonist constraints, strategies, and mobility: recent trends in Latin American frontier zones", in *Land Settlement Policies and Population Redistribution in Developing Countries*, A Oberaie (Ed), Praeger, New York, pp. 271–316.

Foweraker, J 1981, *The Struggle for Land*, Cambridge University Press, Cambridge.

Goodman, D 1977, "Rural structure, surplus mobilization, and modes of production in a peripheral region: the Brazilian Northeast", *Journal of Peasant Studies*, Vol. 5, No. 1, pp. 3–32.

Gould, J 1972, *Economic Growth in History, Survey and Analysis*, Methuen, London.

Gray, H 1976, *A Generalized History of International Trade*, Holmes and Meier, New York.

Green, S and Perlman, S 1985, "Frontiers, boundaries, and open social systems" in *The Archaeology of Frontiers and Boundaries*, S Green and S Perlman (Eds), Academic Press, New York.

Henkel, R 1981, "The move to the Oriente: colonization and environmental impact", in *Modern Day Bolivia: Legacy of the Revolution and Prospects for the Future*, J Ladman (Ed), Center for Latin American Studies, Arizona State University Press, Tempe, pp. 277–300.

Hindess, B and Hirst, P 1975, *Pre-capitalist Modes of Production*, Routledge and Kegan Paul, London and Boston.

Hudson, J, 1969, "A location theory for rural settlement", *Annals of the Association of American Geographers*, Vol. 59, pp. 365–81.

Kristof, L 1959, "The nature of frontiers and boundaries", *Annals of the Association of American Geographers*, Vol. 49, pp. 269–82.

Janeczek, A 2011, "Frontiers and borderlands", *Quaestiones Medii Aevi Novae*, Vol. 16, pp. 5–14.

Johnston, R, Gregory, D, and Smith, D (Eds) 1994, *The Dictionary of Human Geography*, third edition, Blackwell, New York.

Lewis, K 1984, *The American Frontier: An Archaeological Study of Settlement Pattern and Process*, Elsevier, Orlando, FL.

Leyburn, J 1935, *Frontier Folkways*, Yale University Press, New Haven.

Lonsdale, R and Holmes, J 1981, *Settlement Systems in Sparsely Populated Regions: The United States and Australia*, Pergamon, New York.

Marquette, C 2006, "Settler welfare on tropical forest frontiers in Latin America", *Population and Environment*, Vol. 27, Nos. 5–6, pp. 397–444.

Martins, J 1996, "O tempo da fronteira, retorno à controvérsia sobre o tempo histórico da frente de expansão e da frente pioneira", *Tempo social*, Vol. 8 No. 1, pp. 25–70.

Mayhew, S 1997, *Dictionary of Geography*, Oxford University Press, Oxford.

Meyer, D 1980, "A dynamic model of the integration of frontier urban places into the United States system of cities", *Economic Geography*, Vol. 56, pp. 120–40.

Muller, E 1977, "Regional urbanization and the selective growth of towns in North American regions", *Journal of Historical Geography*, Vol. 3, pp. 21–39.

Neiva, AH 1949, "A imigração na política brasileira de povoamento", *Revista Brasileira de Municípios*, Vol. 2, No. 6, pp. 222–44.

Prescott, J 1965, *The Geography of Frontiers and Boundaries*, Aldine, Chicago.

Rey, P 1975, "The lineage of mode of production", *Critique of Anthropology*, Vol. 3, pp. 27–79.

Ruccio, D and Simon, L 1988, "Radical theories of development: Frank, the Modes of Production school and Amin," in *The Political Economy of Development and Underdevelopment*, Wilber, C (Ed), Random House Business Division, New York, pp. 121–73.

Steffen, J 1980, *Comparative Frontiers: A Proposal for Studying the American West*, Oklahoma University Press, Norman.

Taylor, J 1979, *From Modernization to Modes of Production – A Critique of the sociologies of Development and Underdevelopment*, Macmillan, London.

Todaro, M 1989, *Economic Development in the Third World*, Longman, New York.

Turner, F 1893, *The Significance of the Frontier in American History*, State Historical Society of Wisconsin, Madison, WI, viewed 2 April 2018, https://archive.org/details/significanceoffr00turnuoft (accessed 2 April 2018).

Vance, J 1970, *The Merchant's World: The Geography of Wholesaling*, Prentice Hall, Englewood Cliffs.

Wallerstein, I 1974, *The Modern World-System* (Vol. 1), Academic Press, New York.

Wallerstein, I 1980, *The Modern World-System* (Vol. II), Academic Press, New York.

Warf, B (Ed) 2006, *Encyclopedia of Human Geography*, Sage, London.

Weigert, H, Henry, E, John, R, Fernstrom, E and Dudlery, K 1957, *Principles of Political Geography*, Appleton-Century-Crofts, New York.

Wilber, CK and Jameson, KP (Ed), 1992, *The Political Economy of Development and Underdevelopment*, Random House Business Division, New York, pp. 3–27.

3 Policy mobilities and land settlement schemes: the emergence of homesteading in Western Australia

Matthew Tonts
Julia Horsley

Introduction

One of the defining characteristics of Australian land settlement policy is the way in which it has freely borrowed and adapted approaches used in other parts of the world, most notably the United States, Canada and New Zealand. Indeed, during the nineteenth and early twentieth centuries, there were dozens of examples of land settlement "experiments" in Australia that have their origins elsewhere (Roberts 1924; Glynn 1975; Powell 1988). Yet, the processes through which policy ideas moved across time and space are not well understood. In contrast, research into contemporary policy mobilities has identified a complex set of interactions at play, including global knowledge sharing and exchange, local social and political receptiveness and policy adaptation (Bensen and Jordan 2011; Peck 2011; McCann and Ward 2013). Central to this research is interrogating the flows, spaces, subjects and webs of relations that underpin the movement and adaptation of policy ideas (Brenner et al. 2010; Peck and Theodore 2010; McCann 2011; Peck 2011; Bok and Coe 2017).

In this chapter, we trace the emergence of Western Australia's Homestead Act of 1893 as a means of better understanding the processes associated with policy mobility in the late nineteenth and early twentieth centuries. This act, designed to stimulate intensive agricultural development in Western Australia through the provision of land to new settlers under specific terms and conditions, has a long and spatially complex history and is largely a replication of the Canadian Dominion Lands Act of 1872. The Canadian act, in turn, draws on the United States' Homestead Act of 1862, which was influenced by notions of "Jeffersonian democracy" and the ideal of the "yeoman farmer" (Smith 1950). We argue in this chapter that the policy ideas embedded in the Western Australian Homestead Act reflect processes of policy mobility and adaptation that bear a number of distinctive resemblances to contemporary urban and regional policy transfer.

Policy transfer, diffusion, mobilities and mutation

Scholarly interest in how policy ideas move within and across political, legal and administrative jurisdictions became increasingly common from the 1960s and built on the burgeoning research interest in the "diffusion of innovations" (Berry and Berry 1999; Bensen and Jordan 2011). Much of this research focussed on questions of geographic propinquity, timing and communication channels in the movement of policy ideas (Dolowitz and Marsh 1996; Peck and Theodore 2010; Dolowitz and Marsh 2000). Over recent decades, this work has expanded to incorporate an appreciation of the role of the actors and institutions involved in the movement of policy ideas, drawing particular attention to the role of elected officials, civil servants, pressure groups and large national and supra-national institutions (Dolowitz and Marsh 2000; Temenos and McCann 2013).

One of the ongoing criticisms of the policy diffusion literature is that it tended to overlook the spatial and temporal complexity associated with the movement of policy ideas from place to place (Peck 2011). For example, relatively little attention has been paid to political motivations, social interactions and the processes of learning and adaptation that occur in the movement of policy ideas. Moreover, a number of scholars have pointed out that policy transfer involved more than mere "lesson drawing" from other jurisdictions (Rose 1991, 1993, 2005). Lesson drawing implies that political actors or decision makers in one jurisdiction simply draw relevant insights from other places, which they then apply to their own settings. As Dolowitz and Marsh (1996, 2000) point out, however, policy transfer is seldom simply about drawing lessons, rather it is underpinned by competitive pressures. Indeed, they argue that a jurisdiction can feel "pressured" to adopt policy from other geographical locations if political actors perceive their own jurisdiction to be "falling behind" its competitors in some way.

The question of global competition generated by international "consensus" on what constitutes "good" or "effective" policy has become a central focus of recent literature on policy mobilities. This has become particularly evident in urban policy, where strategies are being widely shared and replicated across much of the developed world (McCann and Ward 2013). Often this transfer and adoption is accompanied by discourses emphasising global "best practice" and the need to remain competitive.

In understanding the process of policy transfer, increasing focus has been placed on individual actors or agents (Peck 2011; Cochrane and Ward 2012; Bok and Coe 2017). In earlier policy diffusion studies, the decision-making rules of state policy makers were seen to hold the key to explaining the rate and distribution of policy diffusion, above and beyond issues of ideology or power; policy innovations were seen as little more than inert data units, travelling across a landscape marked only by formal jurisdictional boundaries (Walker 1969; Peck 2011). The emergence of more multidisciplinary perspectives on how, why, where and with what effects policies are mobilised, circulated, learned, reformulated and reassembled has deepened our understanding of policymaking as both a local and, simultaneously, a global socio-spatial and political process (Bensen and Jordan 2011; McCann and Ward 2013).

The focus on social construction, relationality, representation, assemblage, multiscalar practice and politics (broadly defined), in contrast to the traditional rational-choice paradigm of political science, has spurred a rolling conversation on the characteristics of policy mobilities and mutations (Bensen and Jordan 2011; Peck 2011; Temenos and McCann 2013). Through this process, a research agenda has begun to emerge that offers a rich conceptualisation of ongoing practices, institutions and ideas that link global circuits of policy knowledge and local policy practice, politics and actors (e.g., Bulmer and Padgett 2005; McCann 2008; Brenner et al. 2010; Peck and Theodore 2010; Peck 2011; McCann and Ward 2013). This conceptual work not only informs but also benefits from detailed empirical research into how the local, and sometimes immobile or fixed, aspects of a place interact with policies mobilised from elsewhere. Policies are not only local constructions; nor are they entirely extra-local impositions on a locality (McCann and Ward 2013). Rather, policies are relational assemblages of fixed and mobile elements including expertise, regulation, institutional capacities and ideology, deployed from and across multiple scales for particular interests and purposes (Bulmer and Padgett 2005; McCann and Ward 2013; Temenos and McCann 2013).

This increasingly relational and contextual view of policy mobility is particularly relevant for interpreting the ways in which land settlement ideas were shared, adapted and implemented in the nineteenth and early twentieth centuries. In hindsight, it is clear that, in contrast to policy transfer being a function of objectified policy "problems" that called forth appropriate "solutions", such as a need to settle and cultivate more land, for example, policy mobilities also reflect politicised opportunities for intervention. Even early "policy lesson learning", such that it was in evidence, was being normatively pre-filtered. As Robertson (1991, p. 64) argued, "[l]essons from other polities are more likely to be used to attack the *status quo* with evidence that feasible and potentially superior alternatives exist elsewhere". In other words, policy transfers can too easily be portrayed as "ideologically exclusive" while, in reality, lesson-drawing is an intensely political, social and cultural phenomenon (Robertson 1991, p. 64). Tracing and interpreting specific examples of policy mobility at different historical-geographical conjunctures serves to further illuminate the tension between policy as fixed, territorial or place-specific, on the one hand, and dynamic, global and relational on the other (McCann and Ward 2011; Temenos and McCann 2013).

Homesteading in the United States and Canada

The Western Australian Homestead Act can trace its origins to a series of homesteading policies that emerged in the United States and Canada during the middle part of the nineteenth century. In essence, the central objective of homesteading policies was to transfer vast areas of "unproductive" public land into privately owned farms. For LeDuc (1962, p. 223), there were three key features of this form of settlement in North America: (i) transfer of title to individual landholders, (ii) an increase in settlement and the uptake of productive land uses and (iii) economic development through the expansion of agriculture and related industries. In addition, homesteading aimed to bring order to land settlement through the use of standardised survey, ordinance and land allocation methods. Yet, it is also clear that homesteading was not simply a rational land allocation methodology used to foster economic development but was also about the enactment of deeply held cultural and ideological values (Smith 1950). Much of this value set builds on Jeffersonian notions of yeoman agriculture and the way in which small, independent farmers represented the claimed values and virtues that were widely seen as central to the formation of the American republic (Peterson 1960).

The Land Ordinance of 1785 provided the basic settlement framework that underpinned homesteading policies (Billings 2012). Rather than the haphazard method of surveying and setting real estate boundaries that had existed previously, the 1785 Ordinance standardised federal land surveys that used astronomical starting points to divide territory into six-mile townships. Each township was divided into 36 sections of one square mile and fourth (or quarter) sections consisting of 160 acres.

While the Land Ordinance provided a framework for the orderly survey of public lands, there was little agreement over the mechanisms that were most appropriate for allocating land to settlers (Wilkinson 1992). Disputes centred around the appropriate price of land, the size of the parcels that should be offered, the implications of homesteading on the practice of slavery in the South and the potential drift of industrial labour in the North to farming in the American West (Kent Rasmussen 2009).

However, the expansion of the United States' continental territory in the 1840s and 1850s brought about a marked increase in the imperative to settle the western United States (Krall 2001). Considerable division over how this was to be achieved remained, however, and, in 1851, 1852 and 1854, homesteading legislation was passed in the US House of Representatives, only to be rejected in the Senate (Kent Rasmussen 2009). A further attempt in 1860 saw President Buchanan veto the bill. The secession of the southern states from the Union shifted the political dynamics considerably and, in 1862, the Homestead Act was passed. Under the provisions of the US bill, settlers 21 years of age or older who were, or intended to become, citizens and who acted as heads of households could acquire tracts of 160 acres of surveyed public land free of all but $10 registration payments. Titles to the land went to the settlers after five years of continuous residence. Alternatively, after only six months, the claimants could purchase the land for $1.25 per acre. Over the years, amendments and extensions of the act made it applicable to forest land and grazing land and increased the maximum acreage tract that individual settlers could acquire (Mikesell 1960).

Importantly, the Homestead Act was not passed in isolation and, later in 1862, the Pacific Railroad Act was passed in order to improve transport between industrial regions and the nation's newly emerging agricultural areas (Billings 2012). By the 1930s, 285 million acres had been allocated under the Homestead Act. While the overall success of the Act remains open to debate (Billings 2012), it nevertheless played a critical role in the settling of large areas of the western United States. Moreover, as a policy instrument, it has had considerable influence beyond its immediate jurisdiction.

To the north, the newly confederated Dominion of Canada expanded rapidly westward through the purchase of a vast area of land held by the Hudson's Bay Company in 1870 (Gerhard 1959). This area today covers northern Quebec and Labrador, northern and western Ontario and much of Saskatchewan, Manitoba, Nunavut and Alberta. One of the central objectives of the Dominion government was to increase economic development, particularly on the Canadian prairies, through the expansion of agricultural enterprise. In developing these regions, Canada drew heavily on American land settlement ideas (Mackintosh and Joerg 1934). From 1871, the Dominion Lands Survey adopted a modified version of the American square or township and range system, with the newly acquired lands being divided into six-mile-square townships containing 36 sections of 640 acres. Each section was divided into "fourths" comprising 160 acres. For Richtik (1975), the adoption of the American system was partly a matter of convenience, in that it provided a framework that would rapidly and accurately provide the country with an orderly pattern of land settlement. However, he also notes that competition with the United States for settlers was critical given the existence of the 1862 Homestead Act. The prevailing view was that any Canadian system of homesteading needed to at least match the offering south of the border (Richtik 1975). Perhaps not surprisingly, therefore, Canada's response, the Dominion Lands Act, bears striking similarities to its American counterpart.

Like the Homestead Act, the Dominion Lands Act provided 160 acres to settlers who could live on, cultivate and improve the land allocated to them. The Canadian law also had a number of more generous provisions than the US law, including the ability of a settler to purchase, at a low price, the 160 acres adjoining their free parcel and a lower minimum age for eligible males (18 years in Canada compared to 21 in the United States). Despite this, the initial success of the Canadian homesteading policy was modest. Between 1872 and 1896, applications for homestead properties averaged only 3,000 per year, with approximately the same number relinquishing their holdings over the same period (Kent Rasmussen 2009). By contrast, the plains of the United States were filling up with homesteaders over the same period, and an estimated 120,000 of them were emigrants from Canada (Kent Rasmussen 2009).

There are a number of apparent reasons for the low rate of initial uptake of Canadian homestead farms, largely associated with absence of critical transport infrastructure. The US transcontinental railway was completed before the Canadian Pacific Railway, ensuring American homesteaders good access to markets. Moreover, between 1872 and 1879, exclusions prevented Canadian homestead farms from being allocated within close proximity to railway land grants, thereby undermining their economic viability (Martin 1973). Other challenges included the more difficult environmental conditions in Canada when compared to the United States and a system of protectionist tariffs that eliminated American competition, which ultimately allowed Canadian profiteers to overcharge for farming implements, supplies and transportation (Kent Rasmussen 2009).

By the 1890s, however, the situation had begun to change when much of the prime agricultural land on the prairies of the United States was exhausted. Moreover, an increase in railway and road infrastructure, assisted migration schemes and investment in social services resulted in a tremendous increase in immigration by settlers from the United States, western Europe and other parts of the world. By the time the Act was repealed in 1930, it had resulted in grants of more than 478,000 square kilometres (Martin 1973).

The Dominion Lands Act can be interpreted as a political and economic response to Canada's land settlement aspirations in the context of competition with the United States. The significant challenges in expanding agricultural settlement in western Canada required a policy framework that would induce migration and development. Yet, this was occurring immediately after the United States had experimented with, and finally established, a systematic policy framework to foster agrarian settlement. As Richtik (1975) pointed out, Canada faced two significant risks. The first was that it would be unable to compete with the United States for new agrarian settlers and investors. The second was that any lag in the Canadian development of the prairies might result in the United States taking an interest in those areas north of its borders. In this sense, the movement of the basic principles of the Homestead Act north to Canada under the Dominion Lands Act can be interpreted as a form of coercive policy transfer (Peck 2011). That is, rather than a "purely technical" response based on real or perceived domestic conditions, the "transfer" came about through a desire to emulate policy in another country lest Canada be left behind in establishing a settled population and fail to divert the emigration from other parts of Canada and the world.

Homestead policy in Western Australia: the 1893 Homestead Act

The emergence of homesteading policy in Western Australia was the result of a number of interrelated factors. First, the colony had experienced difficulties in establishing viable, large-scale agricultural settlement since its foundation in 1829. Second, it had political leaders with direct exposure to the US and Canadian experiences of agricultural land development and settlement. Third, the colony had a history of being receptive to new ideas and policy approaches to agricultural development. Fourth, the granting of self-government in 1890 increased the level of financial and policy-making autonomy within Western Australia.

The initial development of the Swan River Colony in 1829 (renamed the Colony of Western Australia in 1832) was on the basis of a survey in 1827 that reported on the region's substantial agricultural potential (Crowley 1960). Indeed, the excitement caused by the initial reports led to widespread land speculation and the emergence of a number of overly optimistic private settlement schemes (Cameron 1981). Yet, the reality was that the environmental, economic and social conditions in the colony were difficult. The poor quality of the soil, harsh climatic conditions, the absence of adequate labour supplies, remoteness from suppliers and export markets and the high cost of infrastructure all worked against early agricultural settlement (Battye 1924).

In its initial six decades, Western Australia experimented with numerous land development models, ranging from the settlement schemes and activities of private consortia, to Wakefieldian "systematic colonisation", to convict-led development, to land grant railways (Hasluck 1965; Cameron 1981; Tonts 2002). Yet, the expansion of agriculture occurred in a slow and haphazard manner. Indeed, such was the concern about the ability of a local legislature to manage land that, in 1889, the proposed Western Australian Constitution was delayed in the British Parliament. In response, the then Surveyor General and Commissioner of Crown Lands, John Forrest, prepared a *Report on the Land Policy of Western Australia from 1829 to 1888* (Forrest 1889). The surveyor general provided an account of the modes by which the Crown alienated some five million acres of land. Forrest's report points out that, for much of its colonial history, land regulations were framed without any reference to a local legislature. This meant that there was often a lack of adequate input and understanding of the local context from within the colony. This issue was, arguably, not fully resolved until the granting of self-government in 1890.

In order to stimulate agricultural development, a series of land reforms were enacted during the 1880s that incorporated greater flexibility in pricing and greater clarity on the conditions under which land was granted to settlers (Forrest 1889, p. 4). The most significant reforms were brought together under the Land Regulations of 1887, which had as their central objective the cultivation and improvement of lands and, perhaps equally noteworthy, the prevention of the acquisition of large areas by individuals or private companies. This last component of the land regulations was in part to limit large areas of the colony being held by speculative investors. However, more important was the government's intent to promote small-scale, intensive agriculture.

Critical here is the role of John Forrest who, in addition to being Surveyor General between 1883 and 1890, became Western Australia's first Premier following the granting of self-government in 1890. As Surveyor General, Forrest had long advocated for an approach to land settlement that promoted small-scale, intensive agriculture in those areas where environmental conditions could sustain this type of farming (Crowley 2000). Indeed, Forrest was deeply sceptical of the capacity of large-scale private capital to develop Western Australian agriculture. In a speech made in 1896, Forrest commented:

> If there is one thing more than another that has been proved to the satisfaction of all of us, in regard to these English companies which undertake the settlement of land in this Colony, it is that they seem unable to develop the lands which they hold here. I think their land settlement has been a complete failure ... these lessons have taught us to be very careful in handing over great enterprise to private individuals.
>
> (PDWA 1896, p. 691)

It is likely that these views reflect not only Forrest's observations as an explorer, surveyor and politician but also his personal upbringing. Forrest's parents arrived in Western Australia in 1842 as part of a private Wakefieldian land settlement scheme at Australind, south of present-day Perth (Crowley 2000). Although they were initially engaged as labourers on the scheme, the Forrest family gradually acquired small parcels of their own agricultural land. The Australind scheme struggled to attract settlers and labour, in large part because of the price of land and the difficult environmental conditions relative to other parts of the British Empire. Hasluck (1965) also notes that the Forrests' situation was typical of that of many small settlers in the mid 1800s when limited infrastructure, the high cost of land, social isolation and difficult environmental conditions made farming extremely difficult. According to Crowley (2000), Forrest knew first hand the effects of widespread poverty and distress on the small population of Western Australians, particularly in rural areas.

Forrest's commitment to small-scale, intensive agriculture appears to have been fortified in 1887 when he toured extensively through the United States and Canada. Indeed, as Forrest commented, what he saw in North America greatly influenced his thinking with regard to the economic advantages of federation, the importance of railways and the need to advance agricultural settlement (*West Australian* 1890). As Western Australia's first premier, Forrest was in a position to bring his personal values, professional experience and observations and political influence to bear on a new agenda for land settlement in Western Australia.

Soon after Forrest's tour of North America and his becoming premier, the notion of homesteading began to feature in Western Australian political discourse. For example, in 1890, the *West Australian* newspaper reported on one of Forrest's speeches given in the south-western town of Bunbury where he emphasised the importance of attracting new farmer settlers to the colony. Forrest noted that the government was able to provide good quality land, and that "he knew of no better inducement to immigrants than that which had been offered in the United States of America and Canada" (*West Australian* 1890, p. 3). In describing the US and Canadian homestead policies, he went on to suggest that "the introduction of this scheme would be of great benefit to the colony and would attract population from all parts of the world" (*West Australian* 1890, p. 3).

Forrest's commentary on land settlement during the early 1890s also seemed to reflect his personal values in relation to the importance of small-scale family farming. Indeed, it often included direct reference to Jeffersonian notions of "yeoman farmers" and their associated virtues. In 1893, for example, Forrest famously argued for the establishment of "a bold peasantry, their country's pride, men of small means but strong arms and stout hearts" (PDWA 1893, p. 230). In the face of criticism of the notion of small-scale settlement, Forrest vigorously defended his vision:

> I am aware that some persons laugh at the idea of us having a race of peasant proprietors. They seem to have received the suggestion as a good joke, when I spoke about a bold peasantry. But the expression is a good one, and I am a great believer in a peasant proprietary—I mean people living on small areas of land and possessing them as their freeholds.
>
> (quoted in Crowley 2000, p. 126)

The ideal of a particular cultural landscape based on small-scale agriculture was also represented in the public policy discourse. For example, the 1891 Commission on Agriculture argued that "the object and desire of every good government should be the permanent settlement of the Colony by a yeoman class on lands yielding each its special product according to its capabilities and its uses" (Western Australia, Parliament 1891, p. 13). There is little doubt that this in part reflects Forrest's influence, but it also suggests a broader set of cultural values that were present within the policy community at the time.

In November 1892, Forrest introduced the first Homestead Bill into the Western Australian Parliament. In his second reading speech, the Premier acknowledged the successful establishment of various railways in regions with agricultural potential. However, he noted that "it is no use providing means of transit throughout the length and breadth of the colony unless we have people to settle and occupy and improve the land through which our railways run" (Forrest, quoted in PDWA 1892, p. 72). He went on to note that, notwithstanding these transit facilities, the area of land in crop had not increased in the previous five-year period and that, of all the five million acres of land that had been alienated from the Crown from the beginning of the colony, only a little over 1% was under crop in 1891.

Critically, Forrest also acknowledged the link between his proposed policy and those in the United States and Canada. He stated:

> Hon. members are no doubt all aware that this proposal of the Government is not altogether a new one – it is not one that they have invented – but it is a system by which the occupation and cultivation of the land has been advanced in the great continent of North America. Those who have travelled through that country, or those who are acquainted with its history, must all know that the chief plan of settlement in the United States of America and the Dominion of Canada in the past has been based on the homestead system; that is, a free grant of land to those who are willing to occupy and settle upon the land, on what is called there "a quarter section" of 160 acres.
>
> (Forrest, quoted in PDWA 1892, p. 72)

The Premier then went on to explain that the Western Australian provisions were essentially the same as those in the earlier Canadian Dominion Lands Act. To facilitate the cultivation of land, the Homestead Bill proposed to grant willing settlers 160 acres of land that was otherwise lying "idle and worthless" in return for their undertaking to cultivate it. However, in addition to the US and Canadian homesteading provisions, Forrest's bill provided loan funds to settlers who had suitably improved their properties. Improvements were taken to mean the construction of a habitable house, fencing, clearing and cropping.

Apart from the objective of reducing imports relative to exports of staples that could be produced locally, there was also an explicit goal of making Western Australia more attractive to "people of the right stamp":

> Other countries all over the world are offering attractions to induce people, and especially people who are ready to enter upon the land and to cultivate it, to come to their shores. In fact, as we all know, there is a race between the civilised countries of the world to secure an addition to their population in the shape of persons who are ready to settle upon the soil and to cultivate it. Is it to be wondered at that we here should desire the same thing? We have tried, and are still trying, to provide cheap and easy means of communication throughout this vast territory of ours; but I say that our efforts in this direction will be entirely futile unless we are also able to obtain people to settle upon the land and to cultivate it, and so provide traffic for these railways that we have built.
>
> (Forrest, quoted in PDWA 1892, p. 77)

Forrest's comments also recognised an important challenge facing Western Australia in the 1890s with regard to global competition. As he pointed out, Western Australia was in competition with many other parts of the world in securing settlers to foster agricultural expansion. This included competition with the very countries from which he had appropriated home-steading policy ideas. However, Western Australia had the added disadvantages of even greater isolation from markets, underdeveloped infrastructure and relatively high farm input costs (Glynn 1975). Adding financial support for new settlers through a loan system provided a means of offsetting these disadvantages and enhancing the state's competitiveness.

Forrest's vision for the implementation of homesteading policies did not proceed without considerable debate. Much of the opposition was from larger landholders who contested his claim that the granting of free land, and especially loans, would attract the right "stamp" of settler (Richardson, quoted in PDWA 1892, pp. 85–90). Particular objection by members of Parliament speaking in opposition to the bill was directed at the financial assistance provisions, with the claim made that they would attract "a poor, indigent helpless class of settlers, without experience or fitness for the work" and not the more desirable class of persons with capital (Richardson, quoted in PDWA 1892, p. 89).

In response, Forrest simply emphasised that the scheme proposed by the Homestead Bill was not new and had been successfully implemented in Canada and the United States. Moreover, he noted that the question of government loans to farmers, the development of agricultural resources and the adoption of a policy of small-scale agriculture had been discussed for many years in the colony. By providing provisions for financial assistance, Forrest was addressing the issue that smaller settlers on the land were generally "struggling men who may have little capital and are not men of means" (Wyatt, quoted in PDWA 2012, p. 3). Noting that Western Australia was a long way from the "great centres of population of the old world", effort was necessary to divert some of the stream of immigration away from the "great countries to the westward of England, Canada, the United States and South America" (Forrest, quoted in PDWA 1892, p. 84). The free-grant system was the foundation of settlement on which "those great countries had flourished" (Forrest, quoted in PDWA 1892, p. 84). Ultimately, however, Forrest was forced to withdraw the original version of the Homestead Bill and reintroduce it without the loan assistance provisions. This version of the bill was assented to in November 1893.

The Homestead Act of 1893 provided that homestead farms might be proclaimed within 40 miles of a railway. Any person being the sole head of a family, or being a male who has attained the age of 18 years, and who was not already the owner of land within the colony exceeding in area 100 acres, was able to select a homestead farm of 160 acres in an area set apart for that purpose (Section 4). The selector was required to take possession in person of the homestead farm within six months and had to reside upon it for at least six months during each of the first five years (Section 6). Within the first two years, the selector needed to expend £30 on the erection of a house, or in clearing and cropping (Section 8) or in planting two acres of orchard or vineyard (Section 9). Within the first five years, the selector was required to fence, quarter, clear and crop at least one-eighth, and within the first seven years must clear and crop at least one-quarter, and fence in the whole of the homestead (Section 8).

If these conditions were complied with, the selector became entitled to a grant in fee simple of the homestead farm at the end of seven years (Section 11); non-compliance with the conditions entailed forfeiture (Section 9). Village sites in connection with homestead areas might be proclaimed and were to consist of allotments of one acre each (Section 14). There were also other provisions for settlement on what was regarded as "inferior land" (Sections 18 and 19). Homestead leases were also obtainable and were granted for 30 years, after which, if the required conditions were fulfilled, the lessee obtained a grant in fee simple (Section 25). In Class 2 quality land, the area of such a lease was to be not less than 1,000 nor more than 3,000 acres; in Class 3 quality land, the area was not to be less than 1,000 nor more than 5,000 acres (Section 33).

In addition to providing land for homestead settlement, Forrest remained convinced of the need to offer financial support to new settlers. By the time the Homestead Act was introduced, Western Australia's overall financial position had improved considerably as a result of major gold discoveries at Coolgardie in 1892 and Kalgoorlie in 1893 (Webb 1993). Forrest was able to use the improving economic conditions of the colony to win support for the financial assistance that he had been unable to incorporate into the initial Homestead Act. The enabling legislation was the Agricultural Bank Act of 1894, and, in 1895, the Agricultural Bank of Western Australia was established to provide support to pioneer farmers. The bank made loans of up to £400 to farmers to assist them in fulfilling the conditions of the Homestead Act, including land clearing, fencing, water management and general improvements. In establishing the bank, the Forrest government recognised that the overall risk was relatively low (Crowley 1960). Immigration rates were high on the back of the gold rush, and almost all settlers were taking up areas of prime agricultural land. Moreover, the pairing of the Homestead Act with the Agricultural Bank Act provided a significant point of difference from the homesteading arrangements in the United States and Canada.

When set alongside the US and Canadian homesteading provisions, it is apparent that the Western Australian homesteading policy was adapted in a number of ways (Table 3.1). While there are similarities reflecting the common heritage of such policies, such as the lot size of 160 acres and the requirements for residency and improvements, adaptations to local conditions are evident in terms of the eligibility of applicants; in Western Australia, the added proposal to provide state financial assistance as an extra incentive (partly in recognition of the smaller economy and lack of available markets

Table 3.1 Key characteristics of homesteading policies

Homestead Act 1862 (USA)	Dominion Lands Act 1872 (Canada)	Homestead Act 1893 (WA)
Provision of a free homestead farm of up to 160 acres.	Provision of a free homestead farm of up to 160 acres.	Provision of a free homestead farm of up to 160 acres.
Eligibility restricted to any person who is the head of a family or who has arrived at the age of 21 years and is a citizen of the United States, and who has never borne arms against the United States Government or given aid and comfort to its enemies.	Eligible for head of household or male at least 18 years old. Available to immigrants.	Eligible for head of a family or any male over 18 years of age. Available to immigrants.
Subject to preemption at $1.25, or less, per acre; or 80 acres or less of such unappropriated lands, at $2.50 per acre, which shall not exceed in the aggregate 160 acres.	Selectors required to pay a $10 fee.	Selectors required to pay a £1 fee.
	Residency requirement of 3 years, with at least thirty acres under cultivation, and construction of a permanent dwelling.	Residency requirement of 6 months every year for the first 5 years; dwelling and improvements to be made within 2 years.
		Supported by financial assistance of up to £400 from the Agricultural Bank from 1895.

Source: Adapted from *The Homestead Act of 1862* (USA) (Act of May 20 1862 Stat. 392), *Dominion Lands Act 1872* (Canada), and *Homestead Act 1893* (WA)

in the newer colony). While the original proposal to include financial assistance in the bill itself was rejected, the passing of the Agricultural Bank Act in the following year achieved the same purpose. This "mutation" of the key homesteading policy ideas drawn from the United States and Canada to address local issues in combination with other policy initiatives, such as an agricultural bank, illustrates how even this early form of settlement policy mobility was not merely a transfer of a "fixed unit" across space but rather a dynamic and adaptive process involving local co-constitution of new geographies (Woods 2007; Peck 2011).

The radically improved economic circumstances experienced in Western Australia following the discovery of gold meant that the Homestead Act was able to be adapted further to suit the local environmental and political contexts. Rapid immigration saw the colony's population increase at around 30% per annum during the mid 1890s, and, amongst colonial administrators and politicians, there was a concern that any end to the gold rush would see equally rapid outmigration and economic decline (Webb 1993). Yet it was recognised that there was an opportunity to "capture" and retain mining labour and capital through the expansion of agriculture. In order to hasten the pace of agricultural development, the Land Act of 1898 repealed the Homestead Act and its amendments. Far from signalling a retreat from homesteading policies, the Land Act took a more expansive view of what was required to underpin successful land settlement in Western Australia.

The Land Act of 1898 consolidated the various land statutes and regulations and provided a statutory framework that set out a more formal method for establishing and recording land districts. In the second reading speech, the success of the Homestead Act was acknowledged by the Commissioner of Crown Lands as "one of the most popular provisions of the law", in noting that, by that time, 555 separate homesteads with an aggregate area of 146,000 acres had been taken up (Throssel, quoted in PDWA 1898, p. 631). The new act extended the operation of the provisions to allow the selection of homestead farms to more areas of Crown land (although still largely limited to southern and western districts of the state). While lot sizes remained at 160 acres, more liberal provisions regarding "homestead leases" (renamed as "grazing farms") were added to allow homesteaders to expand their holdings by acquiring second- and third-class grazing land.

The new act also set out provisions for "working man blocks" based on similar mechanisms to the homestead provisions. These blocks were offered on the basis that "the labourer as well as the farmer should be enabled to come straight to... [Western Australia]" and claim their entitlement to land (Throssel, quoted in PDWA 1898, p. 631). These provisions allowed any man who was not already the holder of rural land to apply for a half acre or a maximum of 20 acres within ten miles of a city or town. A lengthy parliamentary debate, reported in Hansard, between the Premier and other members of parliament provides some insight into the cultural values that underpinned this new form of homestead block (Forrest, Illingworth, Venn, Vosper and Kenny quoted in PDWA 1898, pp. 831–49). While there was some disagreement about the size of blocks, their location and the availability of finance from the Agricultural Bank, there was broad consensus that providing rural land to "men of independent means" was important (Forrest, quoted in PDWA 1898, p. 831). Indeed, the sentiment expressed during the debate suggests that there remained a strong desire to create a cultural landscape based on small-scale yeoman farming.

Despite the commitment to small-scale agricultural settlement, in the replication of policy ideas from the Canadian Dominion Lands Act and the US Homestead Act, the Western Australian Homestead Act was built upon a set of environmental, economic and technological assumptions that were disconnected from the reality of the prevailing geographical and temporal contexts. The North American policies were more than 20 years old and very much reflected their economic, technological and geographical origins (Allen 1991). Between the 1860s and early 1870s and the passing of the Western Australian Act in 1893, there had been major improvements in the science underpinning agriculture, advances in farm technology and increasing cost-price pressures at the farm level (Glynn 1975). Moreover, the environmental conditions in Western Australia, particularly soil quality and rainfall, meant that farm productivity was lower than that in parts of the Canadian and American West. Collectively, this meant that 160 acres was increasingly acknowledged as being too small to constitute a viable farming enterprise (Tonts 2002). So, while the transfer of policy ideas from North America to Western Australia had resulted in some adaptation to local context, arguably they were not sufficiently well suited to the economic and environmental realities facing local farmers.

There were also other significant changes between the homesteading policies first being mooted for Western Australia in the early 1890s and those passed in the Land Act in 1898. The most significant of these was the discovery of gold. In addition to fuelling rapid immigration, the increased population and increased wealth from gold created considerable demand for agricultural land (Glynn 1975). At the same time, the Western Australian government was actively seeking to diversify the economy should the gold industry falter. This, together with the increased creditworthiness of the colony, created conditions that enabled the government to increase the pace and scale of development.

One of the key measures introduced in response to the shifting economic, technological and social contexts within which agricultural development was occurring was embedded in Part Five of the Land Act, which offered an expansive version of homesteading. Rather than restricting settlement to farms up to 160 acres, provisions were made for any person over 18 to apply for a surveyed block of up to 1,000 acres at a cost of 10 shillings per acre payable half-yearly at the rate of one-twentieth of the total purchase money per annum or sooner (Sections 55[1] and [2]). There was a requirement that the applicant take up residence within six months and reside on the property for at least six months per year for the first five years (Section 55[4]). The new bill also carried forward an amendment to the Agricultural Bank Act in 1896 (Statute 5) that raised the maximum advance from £400 to £800 and extended the scope of improvements to which funding could be applied to include: fencing, draining, wells of fresh water, reservoirs, buildings and any improvements that, in the opinion of the manager of the bank, increased the agricultural or pastoral capabilities of the land (Department of Lands and Surveys 1934, p. 90).

In fulfilling the objective of rapid agricultural development, the Land Act was accompanied by a number of other critical policy instruments that improved efficiency and profitability in agriculture. In 1905, the Royal Commission on Immigration (Western Australia, Parliament 1905) recommended the construction of an extensive network of railways in those areas with the potential for agricultural development. The commission based its findings not only an assessment of local needs but also on the demonstrated importance of railway development in Canada and the United States as a means of promoting agricultural settlement. The outcome was the development of more than 4,000 kilometres of railway that were constructed between 1905 and the onset of World War I (Glynn 1975).

Collectively, the combination of the Land Act, the Agricultural Bank Act, railway construction and other rural development strategies resulted in a rapid increase in the area under agricultural production. In 1899, there were 218,000 acres of homestead farms, and by 1909, this had increased to 1,273,241 (*Statistical Register of Western Australia* 1899, 1909). In 1899 there were 1,141,988 acres under other forms of conditional purchase arrangements and by 1909 this had increased to 5,220,391 acres (*Statistical Register of Western Australia* 1899, 1909). This agglomeration of land settlement policies, as transferred and modified in the 1890s, played a major role in facilitating the speed and spread of agricultural development in Western Australia. While the settlement of Western Australia would probably have progressed in the absence of this particular set of policy conditions, it is not unreasonable to surmise that such development would have occurred at a much slower rate and in a more haphazard manner. In support of a hypothesised "counterfactual", comparative empirical comparisons could be made with the rates of land settlement in other regions of Australia, where the homesteading policy was not adopted and where the natural geographical endowments and human populations were more favourable to economic development. There is much evidence to suggest that this particular co-constituted agglomeration of policies had a significant and enduring impact on land settlement in Western Australia.

Conclusion

In conceptualising the transfer of the homesteading policy from the United States, via Canada, to Western Australia in the 1890s, a range of potentially useful theorisations in policy transfer and mobilities literature can be made. While the significance of the influence of individual "rational" actors is more contested in recent literature (Peck and Theodore 2010; Peck 2011; Bok and Coe 2017), it is evident, in empirical terms, that the actions of Sir John Forrest were, at least ostensibly, in accord with the notion of deliberate "lesson learning" (Rose 1991; Dolowitz and Marsh 1996) as demonstrated by his specific recounting in parliament of the existence and benefits of homesteading legislation in the United States and Canada when introducing this policy (Forrest quoted in PDWA 1892, p. 72). However, this "policy transfer" did not occur against an empty background. Rather there was a complex range of geographical, institutional and cultural factors that brought about its genealogy, mobility and mutation.

First, in political terms, the ability of Forrest to point to the experience of other countries supported his reform arguments and enabled him to overcome the entrenched status quo values of existing land holders who contested the notion of "giving away free land" (Robertson 1991). Second, the long-established desire for small-scale farming had first been transplanted by English colonists at the beginning of Western Australian land settlement in 1829 (Cameron 1981) and was reinforced by the popularity of homesteading policies in other jurisdictions (Robertson 1991; Wolman 1992). While Canada, the United States and Australia all developed land settlement policies aimed at establishing small homestead farms, their differing historical and geographical circumstances and objectives ensured variations in the development, implementation and outcomes of this policy. Notions such as the "rural idyll" in Britain, which valued the aesthetic landscape (Newby 1979) and the "pastoral ideal" in the United States (Bunce 1994), became linked in different ways to the moral and civic virtues associated with taming the wilderness and cultivating the land (Marx 1964) because they evolved within specific social, economic and geographical contexts (Bunce 1994). A uniquely Australian rural ideal holds that the struggle to settle, farm and manage the natural environment engendered values associated with self reliance, entrepreneurialism and resilience (Dempsey 1992).

Third, land policy adheres both to the agrarian ideal and to the economic foundations and forces of capitalism. This "contradiction" in land policy was already evident in Jefferson's position. Although he promoted the agrarian ideal, he also adhered to the principles of laissez-faire capitalism on the land (Krall 2001). This tension between local social policy concerns and the impact of more global economic structural trends is a key feature that early land settlement policy shares with modern policy mobilities.

Peck (2011) suggests that it is, at the very least, an ironic coincidence that the recent rejuvenation in the policy transfer literature – with its appeals to rational actor models and its deference to global "best practice" – should have occurred during the present heyday of neoliberal expansionism. The measurable surge of interest in these new policy mobilities, and the circulatory systems that maintain them, has expanded in tandem with the particular set of institutional and ideological conditions that we have come to associate with neoliberalism (Ferguson and Gupta 2002; Levi-Faur 2005; Peck 2011). The policy transfer process, for all its trappings of "disembedded" policy development, must therefore be understood as an institutionally produced and embedded phenomenon, the character, causes and consequences of which extend beyond the idealised universe of rational-actor models to conjunctural specificity and network relations (Peck 2011). Policy mobility is but one moment in a wider, transformative process, involving the mutation of policies and policy regimes. These are not merely being transferred over space, their form and effects are transformed as are relational connections across a dynamic socio-institutional landscape. As this paper emphasises, the process of global policy mobility is not new, and much agricultural land settlement policy involved a complex process of political and cultural interactions across time and space. Indeed, the global mobility of policy ideas on homesteading in the late nineteenth and early twentieth centuries was not simply based on "rational choice" and a systematic diffusion of policy innovations but was a process mediated by economic conditions, cultural values, political networks and environmental considerations.

References

Allen, D 1991, "Homesteading and property rights; or, 'How the west was really won'", *The Journal of Law and Economics*, Vol. 34, No. 3, pp. 1–23.

Battye, JS 1924, *Western Australia: A History from Its Discovery to the Inauguration of the Commonwealth*, Clarendon Press, Oxford.

Bensen, D and Jordan, A 2011, "What have we learned from policy transfer research? Dolowitz and Marsh revisited," *Political Studies Review*, Vol. 9, pp. 366–78.

Berry, F and Berry, W 1999, "Innovation and diffusion models in policy research" in P Sabatier (Ed), *Theories of the Policy Process*, Westview, Boulder, CO, pp. 169–200.

Billings, RD 2012, "The Homestead Act, Pacific Railroad Act and Morrill Act," *Northern Kentucky Law Review*, Vol. 39, No. 4, pp. 699–736.

Bok, R and Coe, NM 2017, "Geographies of policy knowledge: the state and corporate dimensions of contemporary policy mobilities," *Cities*, Vol. 63, pp. 51–7.

Brenner, N, Peck, J and Theodore, N 2010, "Variegated neoliberalization: geographies, modalities, pathways," *Global Networks*, Vol. 10, pp. 1–41.

Bulmer, S and Padgett, S 2005, "Policy transfer in the European Union: an institutionalist perspective," *British Journal of Political Science*, Vol. 35, No. 1, pp. 103–26.

Bunce, M 1994, *The Countryside Ideal: Anglo-American Images of Landscape*, Routledge, London.

Cameron, JMR 1981, *Ambition's Fire: The Agricultural Colonisation of Pre-Convict Western Australia*, University of Western Australia Press, Nedlands.

Cochrane, A and Ward, K 2012, "Theme issue: researching policy mobilities: reflections on method," *Environment and Planning A*, Vol. 44, No. 1, pp. 5–51.

Crowley, FK 1960, *Australia's Western Third*, University of Western Australia Press, Nedlands.

Crowley, FK 2000, *Big John Forrest: a founding father of the Commonwealth of Australia*, University of Western Australia Press, Nedlands.

Dempsey, K 1992, "Mateship in country towns," in J. Carroll (Ed), *Intruders in the Bush: The Australian Quest for Identity*, 2nd ed, Oxford University Press, Melbourne, pp. 131–42.

Department of Lands and Surveys 1934, *The Agricultural Bank Royal Commission's Report*, Government of Western Australia, Perth.

Dolowitz, D and Marsh, D 1996, "Who learns what from whom: a review of the policy transfer literature," *Political Studies*, Vol. XLIV, pp. 343–57.

Dolowitz, D and Marsh, D 2000, "Learning from abroad: the role of policy transfer in contemporary policymaking", *Governance*, Vol. 13, pp. 5–24.

Ferguson, J and Gupta, A 2002, "Spatializing states: toward an ethnography of neoliberal governmentality," *American Ethnologist*, Vol. 29, pp. 981–1002.

Forrest, J 1889, *Report on the Land Policy of Western Australia from 1829 to 1888*, Government Printer, Perth. [Presented to the Western Australian Legislative Council, third session of 1889. Copy in State Archives, PR1234.]

Gerhard, D 1959, "The frontier in comparative view," *Comparative Studies in Society and History*, Vol. 1, No. 3, pp. 205–29.

Glynn, S 1975, *Government Policy and Agricultural Development: A Study of the Role of Government in the Development of the Western Australian Wheatbelt*, University of Western Australia Press, Nedlands.

Hasluck, A 1965, *Thomas Peel of Swan River*, Oxford University Press, Melbourne.

Kent Rasmussen, R 2009, *Agriculture in History*, Salem Press, Hackensack, NJ.

Krall, L 2001, "US land policy and the commodification of arid land (1862–1920)", *Journal of Economic Issues*, Vol. 35, No. 3, pp. 657–74.

LeDuc, T 1962, "History and appraisal of U. S. land policy to 1862", *Agricultural History*, Vol. 36, No. 4, pp. 222–4.

Levi-Faur, D 2005, "The global diffusion of regulatory capitalism", *Annals of the American Academy of Political and Social Sciences*, Vol. 598, pp. 12–32.

Mackintosh, W and Joerg, W (Eds) 1934, *Canadian Frontiers of Settlement (8 vols.)*, Macmillan, Toronto.

Martin, C 1973, *Dominion Lands Policy*, McLelland and Stewart, Toronto.

Marx, L 1964, *The Machine in the Garden: Technology and the Pastoral Ideal in America*, Oxford University Press, New York.

McCann, E 2008, "Expertise, Truth, and Urban Policy Mobilities: Global Circuits of Knowledge in the Development of Vancouver, Canada's 'four pillar' Drug Strategy," *Environment and Planning A*, Vol. 40, No.4, pp. 885–904.

McCann, E 2011, "Urban policy mobilities and global circuits of knowledge: towards a research agenda", *Annals of the Association of American Geographers*, Vol. 101, No. 1, pp. 107–30.

McCann, E and Ward, K 2011, *Mobile Urbanism: Cities and Policy-making in the Global Age*, University of Minnesota Press, Minneapolis.

McCann, E and Ward, K 2013, "A multi-disciplinary approach to policy transfer research: geographies, assemblages, mobilities and mutations", *Policy Studies*, Vol. 34, No. 1, pp. 2–18.

Mikesell, MW 1960, "Comparative studies in frontier history", *Annals of the Association of American Geographers*, Vol. 50, No. 1, pp. 62–74.

Newby, H 1979, *Green and Pleasant Land? Social Change in Rural England*, Hutchinson, London.

Parliamentary Debates of Western Australia (PDWA) 1892, Government Printer, Perth.

Parliamentary Debates of Western Australia (PDWA) 1893, Government Printer, Perth.

Parliamentary Debates of Western Australia (PDWA) 1896, Government Printer, Perth.

Parliamentary Debates of Western Australia (PDWA) 1898, Government Printer, Perth.

Parliamentary Debates of Western Australia (PDWA) 2012, Government Printer, Perth.

Peck, J 2011, "Geographies of policy: from transfer-diffusion to mobility-mutation", *Progress in Human Geography*, Vol. 35, No. 6, pp. 773–97.

Peck, J and Theodore, N 2010, "Mobilizing policy: models, methods, and mutations", *Geoforum*, Vol. 41, pp. 169–74.

Peterson, MD 1960, *The Jefferson Image in the American Mind*, University of Virginia Press, Charlottesville.

Powell, JM 1988, *The Historical Geography of Australia: The Restive Fringe*, Cambridge University Press, Cambridge.

Richtik, JM 1975, "The policy framework for settling the Canadian West 1870–1880", *Agricultural History*, Vol. 49, No. 4, pp. 613–28.

Roberts, SH 1924, *History of Australian Land Settlement*, Melbourne University Press, Melbourne.

Robertson, D 1991, "Political conflict and lesson-drawing", *Journal of Public Policy*, Vol. 11, No. 1, 55–78.

Rose, R 1991, "What is lesson-drawing?", *Journal of Public Policy*, Vol. 11, No. 1, pp. 3–33.

Rose, R 1993, *Lesson-Drawing in Public Policy*, Chatham House, Chatham.

Rose, R 2005, *Learning from Comparative Public Policy: A Practical Guide*, Routledge, London.

Smith, HN 1950, *Virgin Land: The American West as Symbol And Myth*, Harvard University Press, Boston.

Statistical Register of Western Australia 1899, Government Printer, Perth.

Statistical Register of Western Australia 1909, Government Printer, Perth.

Temenos, C and McCann, E 2013, "Geographies of policy mobilities", *Geography Compass*, Vol. 7, No. 5, pp. 344–57.

Tonts, M 2002, "State policy and the yeoman ideal: agricultural development in Western Australia, 1890–1914", *Landscape Research*, Vol. 27, No. 1, pp. 103–15.

Walker, JL 1969, "The diffusion of innovations among the American states", *American Political Science Review*, Vol. 63, pp. 880–99.

Webb, M 1993, "John Forrest and the Western Australian goldrushes", *Early Days: Journal of the Royal Western Australian Historical Society*, Vol. 10, No. 5, pp. 473–88.

Western Australia, Parliament 1891, "Final report of the Commission on Agriculture", *Votes and Proceedings of the Western Australian Parliament, Part XII*, Government Printer, Perth.

Western Australia, Parliament 1905, "Report of the Royal Commission on Immigration", *Votes and Proceedings of the Western Australian Parliament*, Government Printer, Perth.

West Australian 22 November 1890, "The Hon. John Forrest's address to the Bunbury electors", p. 3.

Wilkinson, C F 1992, *Crossing the Next Meridian: Land, Water, and the Future of the West*, Island Press, Washington, DC.

Wolman, H 1992, "Understanding cross national policy transfers: the case of Britain and the US", *Governance*, Vol. 5, pp. 27–45.

Woods, M 2007, "Engaging the global countryside: globalization, hybridity and the reconstitution of rural place", *Progress in Human Geography*, Vol. 31, No. 4, pp. 485–507.

4   Jewish farm settlements and the Jewish Colonization Association in Western Canada

John C. Lehr

Introduction

Of the plethora of organizations involved in colonial land settlement schemes in the latter half of the nineteenth and early years of the twentieth centuries, few, if any, had the geographic reach of the Jewish Colonization Association (JCA). It was involved in numerous projects on agricultural frontiers in four continents and was unusual in that its land settlement projects were not founded on the desire to turn a profit. Consequently, it had no interest in land speculation, railway development or in securing commissions paid by governments for recruiting agricultural settlers. Its motives were purely philanthropic, directed at alleviating the desperate social and economic conditions of the Jews of Eastern Europe through promotion of agricultural settlement overseas.

In the late nineteenth century, Jewish society in Western Europe had achieved a measure of tolerance and accommodation with its host societies. Although prejudice and discrimination still abounded, it was far less severe than that experienced by those Jews living in Eastern Europe. Under Austrian administration in Galicia and Romania and Russian governance in Ukraine and Russia, Jews were subject to discriminatory laws that generally forbade them to own land, regulated where they could live and restricted the occupations in which they could engage. In consequence, Jews in Russia lived within the pale of settlement, where they were mostly confined to urban ghettos, eking out a living as peddlers, labourers, small merchants or artisans. Whether in Austrian or Russian territories, they remained always "the other", blamed for a host of socio-economic problems not of their making. To assuage Gentile frustration with high taxation and price increases, pogroms were periodically unleashed against Jewish communities when murder, rape and assaults on Jews were carried out with impunity by Christian mobs. As often as not, the authorities blamed the Jews for inciting the pogrom and levied heavy financial penalties on affected communities (Gitelman 1988).

In 1881, following the assassination of Tsar Alexander II, a series of bloody pogroms swept through Russian cities and towns. Thousands of Jews fled for their lives and sought refuge beyond Russia's borders. In Western Europe, numerous committees were formed with the express purpose of aiding these refugees: in England, the Mansion House Committee proffered aid; and in Montreal, the Young Men's Hebrew Benevolent Association and the Ladies' Hebrew Benevolent Society founded the Emigration Aid Society (Sack 1965, pp. 191–6). These societies aimed to ease the lot of Jewish refugees through financial aid and promotion of emigration to mostly urban safe havens. Agricultural settlement was far from being their focus. When they became involved in agricultural endeavours, it was only as an accidental by-product of their benevolence.

The Western European Jewish intelligentsia, some of whom by the latter half of the nineteenth century had achieved considerable wealth and social standing, saw overseas emigration as the vehicle to free their fellows in Eastern Europe from oppressive regimes. They realized Western European sympathy for masses of indigent Jewish refugees, dependent on charity, would soon erode. They thought it vital that Jewish refugees be dispersed to locales where they could make a living and not attract unwanted attention.

Many advocates of Jewish emigration in the late nineteenth century also believed it was necessary for Jews to create a bond with the land, to destroy the image of Jews as people who lived by trade, with no firm ties to place. Movements such as *Am Olam* (The Eternal People) argued that immigration of Jews to agricultural frontiers would create a new kind of more masculine Jew, one who lived by the sweat of his brow and took ownership of the land in both a physical and spiritual way. Reference was made to the agricultural origins of the Jews in Biblical times when the Jews farmed in the land of Israel. To these 'back to the land' advocates geographic locations were immaterial; some of the earliest Jewish colonies were established in the United States, mostly in the West (Shpall 1950, pp. 120–46; Fogarty 1975, p. 157). The Zionist philosophy had yet to be articulated; even after the First Zionist Congress in 1892, proponents such as Moses Hess and Theodore Herzl remained outside of mainstream Jewish thought for years. Thus, the back to the land movement was unconcerned with the location of the lands for Jewish settlement; it was a spatially neutral philosophy.

In the nineteenth and the early twentieth centuries, there were many frontiers open for Jewish colonization: Argentina, Brazil, Australia, New Zealand, Canada, the United States and South Africa (Robinson 1912). Palestine was not considered as a viable emigration destination for Jews. It had the disadvantage of an established agricultural population, and it lay within the bounds of the Ottoman Empire, presenting political obstacles for emigration organizers who also had to contend with environmental and economic hurdles.

Among those who wished to alleviate the plight of Russian Jews was Baron Maurice de Hirsch, a wealthy German Jewish financier. In 1887,

when Hirsch lost his only son, he adopted the Jewish people as his beneficiaries, supporting charities bent on relieving the misery of the Jews of
Eastern Europe. Hirsch was initially unconcerned with Jewish agricultural
settlement overseas but came to realize it was a solution to a growing problem as indigent Jews crowded into the port cities of Western Europe. Keenly
aware that a few Jewish refugees would initially receive a good deal of sympathy, he believed masses of penurious Jews concentrated in a few cities
would quickly erode any goodwill if they remained dependent on public
charity. Hirsch, and others supporting Jewish relief, came to see overseas
emigration as a way of dispersing refugees and giving them a new start.

In 1891, Hirsch set up a fund in New York with capital of $2.0 million
for the education and relief of Hebrew immigrants and their children from
Russia and Romania. Shortly afterwards, he established the JCA with an
initial grant of $2.4 million specifically to initiate Jewish agricultural set-
tlement. Hirsch believed in the regenerative power of the soil and hoped
to establish Jews as farmers and have them move beyond their traditional
occupations in Europe in a complete "from the bottom up" reorganization
of Jewish life as it existed in Eastern Europe. (Tulchinsky 1992, p. 120). To
this end, he insisted loans were to be made only to intending farmers; to pay
for transportation from ports of arrival to places where land or employment
was to be had; for training in mechanics, handicrafts and trades; and for
instruction in the English language, citizenship responsibilities and agriculture (Norman 1985, p. 18). He did not believe in "unproductive charity",
which he thought counterproductive, stifling initiative, breeding a culture of
dependency and eroding self-reliance (Zablotsky 2004, pp. 2–7). This opinion was later borne out when, in Argentina, the JCA provided a cash dole
to some penurious settlers who eventually became dependent upon it and
refused to work (Schwartz and Te Velde 1939, p. 191). Thereafter, aid was
disbursed by the JCA only as repayable loans with interest at five percent
per annum (Zablotsky 2004, pp. 2–7).

Hirsch envisioned a carefully regulated exodus of Jewish emigrants from
Russia, but events overtook him. He planned to buy land in Argentina, sub-
divide it and set up the basic elements – houses, barns, wells and so forth – before
the first settlers arrived. Argentina appealed to him because it was more
sparsely settled than other frontiers then open to settlement and offered
more opportunities for the creation of Jewish farm colonies (Belkin 1966,
p. 68; Avni 1991, pp. 37–9). The first groups arrived in Argentina after land
had been acquired but little else had been done. The preparations made to
receive them were inadequate. Many hired as administrators were "enlight-
ened" westernized Jews who spoke little or no Russian or Yiddish and who
tended to look down on the orthodox immigrants from Russia and Galicia.
Hersch's administrators were impatient with demands for kosher food
and observance of the Sabbath when they impeded progress (Avni 1991,
pp. 31–2). Hirsch wished to have the final say in all decisions, but he had no
appreciation of the difficulties being faced thousands of miles away on the

Argentinian pampas. Chaos ensued. In some areas, settlers rioted in protest over the JCA's ineptitude in supplying their needs (Avni 1991, pp. 15–42). It was not a happy beginning.

Settlement on the prairies: the administrative framework

In Canada, the JCA fared somewhat better. Learning from its experience in Argentina, the JCA adopted a different model for its settlement work, foregoing purchase of land and instead letting settlers acquire land through the "free" homestead system.

Settlement on the prairies was governed by the Dominion Lands Act of 1872. Under the terms of the act, land was surveyed into townships comprised of 36 sections each of one square mile. Each section was subdivided into four quarter-sections of 160 acres. To compensate the Hudson's Bay Company for ceding Rupertsland to the Crown, it was given one and three-quarters of a section in every township; a further two sections were reserved as school lands. In many areas, every odd-numbered section was granted to the Canadian Pacific Railway Company as payment for building the transcontinental railway (Martin 1973, p. 18). This meant that of the 144 quarters in every township, only 96 were available as homesteads (Figure 4.1). The effect of this was increased by regulations that governed the granting of homesteads to settlers.

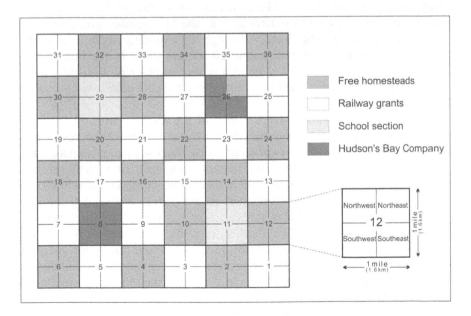

Figure 4.1 Plan of a typical township in Western Canada.

Source: Township Registers

The Dominion Lands Act contained certain provisions aimed at preventing land speculation. After a settler paid a ten-dollar entry fee, he had to reside on the homestead for six months a year for at least three consecutive years, build a habitable dwelling, cultivate the land "to a reasonable extent" and, after 1908, become a Canadian citizen before patent (full title) was granted. "A reasonable extent" of cultivation was usually taken to mean 30 acres, although homestead inspectors took the nature of the land into consideration, often noting that clearing one acre of bush country was equivalent to clearing several acres of "good prairie land." An alternative path to patent was to perform one year of continuous residence, clear and break 30 acres, build a habitable house and purchase the land from the government at the government price, which around the turn of the century ran at upwards of two dollars an acre for land in the parkland belt. Only when patent was granted could the land be sold, mortgaged or rented out (Martin 1973, pp. 137–56; Lambrecht 1991, pp. 22–3). These residency requirements, combined with the reservation of land for schools, the Hudson's Bay Company and the railway companies, mandated dispersed settlement, making it exceptionally difficult to reconcile agricultural settlement with nucleated settlement.

If the financial and political will was present, there was a way to create nucleated settlement within the framework of the Dominion Lands Act. Clause 37 of the act (usually termed the Hamlet Clause) permitted the recognition of hamlets at the discretion of the Minister of the Interior, in which case settlers could fulfil their homestead obligations while living in a hamlet away from their homestead. The clause was not granted lightly or often; and it was clearly aimed at settlers such as the Mennonites who desired to practice cooperative agriculture. Although not explicitly articulated in the act, officials usually expected around 20 settlers to be in the hamlet and some institutional buildings – such as a school or place of worship – to be erected; in other words, there had to be demonstrable elements of a social community.

Officials of the Canadian Department of the Interior and, often, those of the JCA, all ranked economic efficiency more highly than social and religious convenience. Not surprisingly, dispersed settlement became the norm on the prairies. Furthermore, in contrast to its actions in Argentina, in Canada, the JCA was unable to secure a land grant in the realm of 40,000 acres from the government "sufficient in area to locate ... 250 families thereon" (Lousada 1906). Later, it became increasingly reluctant to become involved in the purchase of land either for resale or to rent to the settlers it sponsored. Had it done so, it could have purchased lands not open for homesteading and resold to Jewish settlers, enabling the creation of denser settlements. This would have ameliorated some of the religious issues faced by observant Jews on the agricultural frontier.

In Canada, JCA efforts were largely confined to providing intending Jewish colonists with financial assistance, enough to purchase "a few implements, a couple of horses, a plough, [and] a cow ..." with enough left to buy some household necessities (Tennenhouse 1972). True to Hirsch's philosophy of philanthropy, aid was always in the form of a loan at five percent annual interest. These were given out with great parsimony, secured against the homesteader's chattels and land. When, and if, a recipient of aid managed to fulfill all the requirements of the Dominion Lands Act and applied for full title (patent) to the land, if there were loans still outstanding to the JCA, it petitioned the regional Dominion Lands Office for the patent to be issued in its name. It was released to the owner only when all outstanding debts were cleared (AM Records of Homestead Entry).

It is difficult to gain a clear picture of the general indebtedness of colonists who received loans from the JCA, though JCA annual surveys of the colonies indicate most loans did not exceed the collateral colonists provided in the form of land, buildings or implements. Before 1919, this generally meant loans did not exceed several hundred dollars. Jewish settlers also applied for aid from the Department of the Interior, which gave loans for the purchase of seed grain and provisions, also at the rate of five percent per annum. The department also withheld patent until all its loans outstanding against the property were cleared, but it permitted repayment in kind, usually grain, at an equivalency rate specified when the loan was granted (AM Records of Homestead Entry, Twp. 24, Rge. 6 WPM). For settlers for whom cash was a scarce commodity, this offered a significant benefit above that of JCA aid.

In a 1931 survey of Jewish farmers in Western Canada, Louis Rosenberg (1939, pp. 241–3) noted Jewish farmers had net assets of between $4.95 and $11.00 an acre, considerably less than the average for non-Jewish farmers at $23.27 an acre. Certainly, they remained undercapitalized, lagging in terms of ownership of tractors, automobiles and all kinds of agricultural machinery. While this suggests vulnerability to foreclosure by lenders, it remains unclear as to the extent to which they may have defaulted on JCA loans.

Jewish settlement on the prairies

Jews, unlike some other groups, such as the Ukrainians, showed no preference for any land type (Figure 4.2). Jews settled from the northern reaches of the aspen-parkland belt to the semi-arid grasslands of the southern prairies within the area that Captain John Palliser had declared unfit for agricultural settlement (Spry 1959 pp. 149–84; Gutkin 1980, p. 56). Regardless of location, while some Jewish farm colonies enjoyed initial success, all eventually atrophied, and their populations drifted into the towns and cities of the West (Rosenberg 1939, pp. 223–5). By the end of the twentieth century, only a few remnants of the Jewish farming communities remained (Arnold 1968, 1972; Leonoff 1982, 1984; Katz and Lehr 1991, 1993). Most settlements were completely abandoned; only a handful of individual farms, physically isolated from mainstream Jewish life, some cemeteries and a few synagogues, most disused and deconsecrated, remained as reminders that, in the early

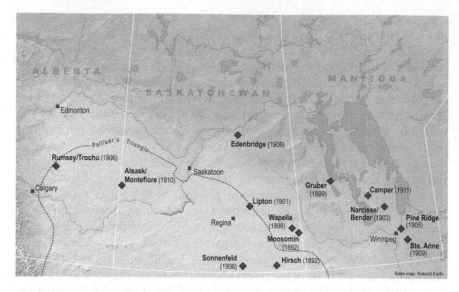

Figure 4.2 Jewish agricultural settlements in Western Canada before 1914.

Source: JCA Reports; AM Records of Homestead Entry; Rosenberg 1939

1920s, there were hundreds of Jews farming more than 500 square miles of Western Canada's prairie (Rosenberg 1939, p. 224).

Jewish settlement endeavours always suffered from systemic problems spawned by the impossibility of reconciling the requirements of Jewish religious observance with the requirements of the Dominion Lands Act. This was compounded by the bureaucratic structure and culture of the JCA, its financial philosophy and its cautious approach to financial aid. A further issue was the JCA's reluctance to engage in the initiation and planning of settlements and its failure to screen settlers heading for Jewish colonies. This attitude, which was in perfect harmony with the prevailing laissez faire philosophy of the time, eroded whatever control the JCA had over the development of Jewish settlements on the prairies.

In Canada, most Jewish farm colonies were established by independent groups of Jewish pioneers. Some were idealists who saw in pioneer settlement the path to attaining their political objectives; others were pragmatists who saw agriculture as a way of making a living or entrepreneurs who hoped that land ownership would bring prosperity. The JCA remained aloof from these varied motivations and regarded pioneer settlement simply as a way to create places for Jews to live (Robinson 1912).

Until 1900, philanthropic work to aid intending settlers was administered by the JCA headquarters in Paris. Distance from the field of operations carried attendant bureaucratic inefficiencies (Belkin 1966, pp. 66–72). The JCA did not become directly involved in agricultural settlement schemes in Canada until the settlement of Western Canada was well under way, even though Jews were among the earliest ethno-religious groups to attempt settlement on the prairies. As early as 1882, partly in response to the "back to the land movement", a scheme to settle Jews at Moosomin (in what is now Saskatchewan) was conceived by the Mansion House Committee of London and its Canadian agent, Sir Alexander Galt. Some 27 homesteads were entered by Jewish immigrants (Rosenberg 1970, pp. 81–2), but harvest failures, remoteness, lack of agricultural experience and undercapitalization combined to ensure its abandonment within a few years (Belkin 1966, pp. 56–7; Tulchinsky 1992, pp. 116–20). A second attempt by Jewish immigrants to settle some 50 miles away at Wapella in 1888 was more successful. It became one of the longest-lived farm settlements in Canada, although it too eventually atrophied in the 1930s, succumbing to the same problems that beset all Jewish farm colonies (Rosenberg 1970, p. 82).

There were early attempts to set up Jewish farm villages but none by the JCA, which became involved in one only after its establishment. In 1899, Rabbi Leiser Gruber established a small hamlet a few miles south of Winnipegosis. The settlement endured only a few years. Although the hamlet he named Gruber was established in an unpromising area of "burnt scrub and numerous swamp", it failed largely through Gruber's ineptitude in dealing with the bureaucracy of the Department of the Interior or possibly because Gruber was more interested in his own financial success than that of his colonists. Regardless, Gruber failed to apply for the Hamlet Clause nor did he obtain permission for his settlers to fulfill their residency requirements while living away from their homesteads. To compound matters, shortly after its inception, Rabbi Gruber abandoned the colony to return to Winnipeg, leaving his settlers to fend for themselves. After considerable negotiation, in 1904, Gruber was recognized as a hamlet and settlers were able to fulfill their homestead residency requirement while living there (AM. Records of Homestead Entry Twp. 30 Rge. 18 W1). The experience so discouraged some that they abandoned their farms to move back to Winnipeg. After Gruber's post office closed in 1907, only a handful of Jewish farmers remained, most of whom left before the end of the decade (AM. Records of Homestead Entry. Twps. 30–31 Rges. 18–19 WPM).

The JCA played no role whatsoever in the short life of Gruber, although within a few years it became heavily involved in the provision of financial aid to settlers involved in a similar scheme. In 1902, Jacob Bender, a land speculator and real estate salesman in Winnipeg, conceived the idea of establishing a settlement of Jewish farmers in Manitoba's Interlake region, which was then in the process of being settled, mostly by Ukrainians and Icelanders. Originally from Nikolayev in Ukraine, Bender recruited his settlers from that region and from the Jewish immigrant population in London, England. As did Rabbi Gruber, Bender envisaged a European-style village settlement that would replicate the social environment of the Jewish shtetl, permitting settlers to live within a Jewish environment yet remain farmers (Richtik and Hutch 1977, pp. 32–3).

Bender acquired a quarter section of land from the Canadian Northern Railway in an area not then served by it. He subdivided his quarter into 19 lots that were sold to prospective settlers, each of whom filed for a homestead on a nearby quarter section. Initially, Bender was oblivious to the need to obtain the hamlet privilege, but he rectified things long before any problems were encountered. Officials of the Department of the Interior felt that, since he and his settlers had acted in good faith and had invested so much time and money in erecting their houses in Bender Hamlet, to reject his application would impose unnecessary hardship. Consequently, the birth of Bender was not attended by the same degree of confusion as that at Gruber. The hamlet prospered until the early 1920s when stock prices fell, and a few families returned to Winnipeg. As people left, remaining in Bender became progressively less attractive and rates of attrition accelerated. By 1925, the settlement was completely abandoned (Rosenberg 1939, p. 224).

Some 20 miles north of Winnipeg, a small Jewish colony was established at Pine Ridge in 1905. Ukrainian and Polish settlers had entered the area almost ten years earlier but, unusually, bought small holdings of 20 to 40 acres from the Metis and Canadian settlers already in the area. Jewish settlers could purchase holdings comparable in terms of size, land quality and accessibility. Pine Ridge had the advantage of being close to Winnipeg. Nevertheless, despite the advancement of loans by the JCA beginning in 1911 and its locational advantage, Pine Ridge's trajectory of decline in the 1920s was little different from those of Jewish colonies elsewhere.

For a brief period (1900–1907), the Jewish Agricultural Society of New York supervised Jewish colonization in Canada. The Hirsch colony, where 73 families were established on homestead land, was established under its auspices. Settlers were provided with houses, livestock, farming implements and provisions for three years for a total expenditure of $50,000. This aid was channeled through the Young Men's Hebrew Benevolent Association of Montreal, which acted as the JCA's agent in Canada at the time (Robinson 1912, pp. 50–1). It was only in 1907 that the JCA created its Canadian Committee, giving a measure of more direct "local" administration, albeit one based in Montreal – more than a thousand kilometres from the nearest Jewish colony in the West. Despite the infusion of aid, the Hirsch colony was no more able to retain its settlers than other, less well-funded, Jewish colonization enterprises.

The Jewish farm colony at Trochu and Rumsey, Alberta, typified Jewish prairie colonies (Figure 4.3). It was founded in 1906 by a small group of Jewish idealists then living in Calgary and had the advantage of being established on good prairie land. It grew to be the largest Jewish colony in the West. By 1911, it had 89 families, 238 people, who cultivated 19,520 acres (Robinson 1912, p. 52; Sanders 1988, p. 2). In 1920, more than 10,000 acres were under cultivation (Stein 2001, p. 13). Like other Jewish colonies such as Lipton/Cupar and Edenbridge, it sprawled over several townships, with clusters of homesteads around Trochu and Rumsey, some 20 miles apart (Figure 4.3).

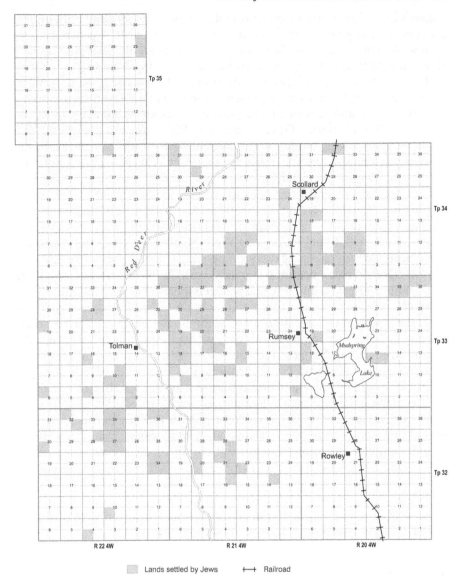

Figure 4.3 Jewish settlement at Trochu and Rumsey, Alberta.

Source: JCA Reports; Sanders 1968

The JCA became involved only in response to requests for financial assistance from settlers already established there. As in other colonies, the JCA funded the provision of Hebrew and Yiddish instruction after regular school hours as well as providing loans to farmers (Sanders 1988, p. 4).

Lipton Colony in Saskatchewan was the only Jewish colony where the JCA attempted something approaching management of its affairs from the colony's inception. Settlers were screened, although not for prior agricultural experience, and given more financial support than the JCA generally gave to Jewish farmers. True to form, the JCA sought an arms-length relationship rather than appointing its own officials and giving them fiscal authority and making them responsible for establishing and supervising the settlement. Instead, JCA representatives approached the Canadian government authorities in England and concluded an agreement to establish a colony under the auspices of the Dominion Government. The JCA agreed to pay for the colonists' transportation but left all arrangements for their placement on the land in the hands of the Canadian Government. The JCA also committed to deposit $200 for each family for the purchase of livestock and farming equipment, entrusting the management of the colony to the Department of the Interior (Belkin 1966, pp. 75–6). JH MacDonald, a local merchant and banker considered to have the necessary familiarity with local conditions and experience in prairie agriculture, was hired. MacDonald chose the site for the settlement and then delegated two local people to run the colony. Neither they nor MacDonald had a common language with the Jewish settlers, and there was "mutual suspicion and mistrust between them" (Friedgut 2007, p. 392).

MacDonald chose to locate the colony in an isolated area, nearly 50 km from the closest railway station. Poor communications meant few markets for any produce. To compound things, settlers were widely scattered; the colony stretched 40 km north to south and 30 km east to west. This was probably quite deliberate. The government was wary of the development of solid blocks of any "foreign" group and so pursued a policy of balancing settlers' desires for propinquity against political fears of alien blocks resistant to assimilation. It was also reluctant to have Eastern European settlers placed in areas close to Anglo-Canadian urban centers where they would attract the ire of the opposition press (Lehr 1977). The result was predictable. Denied any opportunity to achieve dense settlement, Jews could not fully develop their communities. Many early Jewish colonists, discouraged, moved on. Newcomers replaced them, and the colony remained at about the same level of population until the collapse of livestock and grain prices in the 1920s, after which, it slowly atrophied.

The failure of the JCA in Western Canada

Some systemic shortcomings of the JCA in Western Canada have been well documented (Katz and Lehr 1993). In many regards, the problems that afflicted JCA-sponsored schemes in Argentina continued in Canada. Paris remained the head office, separated by time, distance and culture from the areas where its field agents were working. The Paris office strove to retain control of all financial matters, administering loans without any understanding of the conditions faced by those on the agricultural frontier of Western Canada and thus impeding the efficient administration of aid. On one occasion, an applicant from Bender Hamlet waited for two years before receiving a response to his request for aid from the JCA. Although in dire need of assistance, his request was turned down. Without hope, he committed suicide (*Israelite Press* 1923).

Seemingly oblivious to the need for an effective long-term strategy for its agricultural aid programs, the JCA was nevertheless keenly aware that most prospective Jewish settlers were urban, with experience only in the trades or merchandising. In some areas of Russia, Jews had been allowed to purchase small plots of land of up to 30 acres and had taken up farming as an occupation, but they were only a small minority. To better prepare those from an urban background for entry into agriculture, the JCA operated training farms where intending emigrants could receive up to four years of instruction in agriculture. One was at Slobodka Lesna in Galicia, established in 1899; six were in Russia, with others in Turkey and Palestine (Robinson 1912, p. 50). Louis Rosenberg (1939, pp. 229–31) estimated some 31 percent of Jewish farmers in Canada had some prior connection with agriculture, though this estimate included many second-generation settlers, either born in Canada or who spent part of their childhood on a homestead, who later took up farming on their own behalf. Sonnenfeld colony in Saskatchewan was established in 1906 by former pupils of the JCA's agricultural school at Slobodka Lesna, most of whom had worked on farms in Western Canada for some years prior (Belkin 1931). There is general agreement, however, that inexperience in agriculture was not the principal reason that Jewish settlers left the land in the interwar period (Feldman 1982; Katz and Lehr 1991, 1993; Feldman 2003; Lehr 2009; Friedgut 2007, 2009).

While some JCA agents, such as Louis Rosenberg, eventually came to have an intimate knowledge of the reality of agriculture in Western Canada, they learned on the job. In the early years of JCA's involvement in Western Canada, its operatives demonstrated their inexperience and unfamiliarity with the nuances of the system of survey and the Dominion Lands Act. When conducting surveys of the Pine Ridge colony in Manitoba, for example, JCA officials recorded locations as "Pine Ridge, Saskatchewan". In one instance, they reversed the township and range, placing a settler's location far away from the colony; in another case, holdings in Pine Ridge were incorrectly listed as west rather than east of the principal meridian. In other areas where Jews were in the process of proving up on their homesteads and were applying for loans from the JCA, its agents consistently valued the settlers' land at $1000, seemingly unaware that, until the settler had fulfilled all homestead duties and had the patent, the government held the title. Only then could the land be mortgaged, sold or given as security for a loan to a party other than the Dominion Government.

Religious barriers to dispersed agricultural settlement

Most Jewish immigrants from Eastern Europe and Russia into Canada before 1914 were religiously observant. They observed the rules of Kashrut, that is, they kept kosher, attended the synagogue on the Sabbath, gave their children a Jewish education and took care that their children did not marry outside the faith. Some of the lesser-known rules of observance were to have a dramatic effect on Jewish life in rural areas. For example, according to the laws of Kashrut, it is not permitted to ride in a conveyance or ride a horse on the Sabbath, nor is it permitted to walk more than a mile from one's house or to carry anything while doing so. Normal sexual relations between a married couple are not possible without access to a *miqvah* (a ritual bath), and circumcisions, carried out shortly after birth, require the services of a *mohel*. To be kosher, meat must come from an animal slaughtered in a specific fashion by a *shochet*. This impeded the exploitation of wild game that, for non-Jewish settlers, was often a valuable addition to their diet. Some animals cannot be eaten under any circumstances (e.g., rabbits and hares); even those that are permitted under the dietary laws (e.g., deer) cannot be shot but must be trapped and slaughtered in the approved way by a *shochet*. Even for approved animals, slaughtered according to the laws of Kashrut, certain organs and the blood cannot be used, further restricting the settlers' ability to exploit the land's wildlife resources.

A side effect of Jewish dietary laws was discouragement of social interaction between Gentiles and Jews because no observant Jew could eat in a non-kosher household. To some Gentiles, unaware of the rules of Kashrut, such behavior was easily misconstrued. Thus, the opportunities for social interaction were reduced and information from neighbouring, more experienced, Gentile farmers was less easily obtained. This was no trivial matter. Promoters of agricultural settlement frequently advised newcomers to settle close to those who had some experience in western agriculture. Dr. Iosyf Oleskiv (1895), for example, an early advocate of Ukrainian settlement, put it very bluntly, instructing Ukrainian immigrants not to be "know-alls" and to take advice from longer-established settlers.

More serious, though, was the requirement for a *minyan*, a quorum of ten adult males, for the Torah to be read. When settlers were scattered on widely dispersed farmsteads, it was difficult to achieve a minyan. Travel by horse or conveyance was not permitted on the Sabbath, and travel by foot was restricted to 2,000 cubits (about one kilometre) from one's residence, placing the synagogue beyond the reach of most observant Jews living on scattered homesteads.

These requirements affected Jewish settlements established on homestead land where settlers were required to live on an individual homestead for three years. The JCA was keenly aware of the issues facing observant Jews living in farming communities away from the centres of Jewish life, but without a radical policy shift in its settlement strategy, there was little it could do. It funded the building of synagogues where Jews could worship and schools where their children could receive a Jewish education. Admirable though this was, it failed to tackle the issues underlying the incompatibility of adherence to religious orthodoxy and the dispersed system of settlement mandated by the Dominion Lands Act.

To some degree, the settlers in the small Jewish colony at Pine Ridge, Manitoba, were advantaged in that they purchased land and therefore were spared from the residency requirements of the Dominion Lands Act. Their proximity to Winnipeg's large Jewish community meant they could more easily obtain kosher meat, access the services of a *shohet* and *mohel* and visit a *miqvah*. The difficulty of ensuring a quorum for reading the Torah scroll on the Sabbath remained because it was not always easy to assemble ten adult males within the prescribed distance of the synagogue, even though they were able to achieve a far denser settlement than in areas where land was obtained by homesteading.

The lure of the city

As the major centre of Jewish life in Western Canada, Winnipeg offered a chance to live in a familiar Jewish milieu, an opportunity for a Jewish education for their children and a reduction in the chance of intermarriage with the Gentile community. It was a very attractive option for a Jewish family struggling on a smallholding in the bush country, especially when a return to Winnipeg would make possible a return to their former occupations as merchants, peddlers, drivers or artisans in various trades. Jews in Pine Ridge saw Winnipeg very differently from their Slavic neighbours for whom urban life for decades remained an unfamiliar environment, one that offered little more than an opportunity for laboring jobs, whereas ownership of land retained its great allure.

The process of land selection by Jewish settlers differed in some key respects from that of settlers from other backgrounds. Most Jews who decided to try their hand at farming in Western Canada had spent some years in Canada's cities; very few gave their homestead location as their place of residence when they applied for entry on to their homestead. Urban addresses predominated. This stands in sharp contrast to equally impoverished Ukrainian settlers who were taking homesteads in similar areas at approximately the same time. In south-eastern Manitoba, for example, at the time they made formal application to register their homestead entry, virtually all Ukrainian entrants gave the quarter they were applying to enter as their place of residence.

Jewish homesteaders spent their time on their land differently from settlers of many other groups. They were, it seems, more anxious to leave a subsistence economy in favour of a commercial cash economy. Entry into a cash-based economy for a settler on marginal land was difficult. It required significant reserves of capital that few had, and over-extension was a major risk. Jewish settlers could apply for loans from the JCA, but most sought to generate capital through off-farm work. This generally meant retreating to Winnipeg in the winter months to seek employment in their former occupations. In Winnipeg, they had a supportive Jewish community and, most probably, family connections.

Slavic settlers, like the Jews, frequently lacked capital, and like Jews, left their homesteads to seek work on railway construction gangs, on harvesting crews or as general farm labourers across the prairies and northern Great Plains. But they did so during the summer months when they left their families on their homestead to continue the task of clearing and breaking land and maintained continuous occupation. They returned to their families at the onset of winter to cut timber for cordwood and ship it to the nearest rail point. Their homestead was their home. They had no other place to which they could retreat, which was a far different mindset from that of Jewish settlers who felt, albeit subconsciously, they had an urban safety net. There are indications some Winnipeg Jews saw economic opportunity in agriculture as a short-term venture. Acquisition of land was the primary goal, agriculture was the vehicle and residency was the cost. The latter could be minimized if circumstances permitted and the farmer avoided raising stock, as was the case with some farmers in Manitoba's Pine Ridge colony.

As were all prairie farmers, those in the Jewish colonies were affected by global economic conditions. The First World War created ready markets for western agricultural products. Stock and grain prices rose and remained high. Many farmers took advantage of the opportunity to expand their operations and to invest in agricultural equipment. Most Jewish colonies reached their peak around 1918 when synagogues were built, and the foundations of Jewish life had been laid. In the early 1920s, prices fell and many farmers – not only Jewish farmers – found themselves overextended and carrying a heavy burden of debt. As early as 1900, the *Jewish Times* cautioned that Jewish farmers were too concerned with modernization and took on heavy debt to buy "unnecessary implements at up to 12 percent interest" (*Jewish Times* 6 July 1900, p. 6). Later, when agricultural prices collapsed in the early 1920s, they sold out and moved to the city to pursue other occupations. Those who left the colonies were replaced by a new wave of Jewish refugees from Europe, but depressed farming conditions, drought and hail persuaded these newcomers to migrate to the cities as soon as immigration rulings permitted. Even those who hung on found their children leaving for the city to secure marriage partners, pursue higher education and follow careers in the professions. Adults who remained would eventually move to the city to retire close to their children (Sanders 1988, pp. 7–9). Agricultural mechanization, the expansion of agricultural holdings and growing urban opportunities continued to fuel the drift to the cities.

Conclusion

The JCA was heavily involved in most, but not all, Jewish agricultural colonization schemes in Western Canada during the European colonial era. Paradoxically it played, at best, a minor role in shaping the geography of Jewish settlement on the prairies, and it never initiated and carried through a settlement project there. The JCA's mission in Canada was ill defined. The broad terms of its role were clear: to assist Jewish emigrants enter farming as an occupation and to alleviate the plight of Jewish refugees from Europe. It seems the JCA never developed an approach squarely focused on creating Jewish rural communities into which Jewish immigrants could be inserted. Placing immigrants prepared to enter farming directly onto the land and backing them with adequate financial aid would have been a more effective use of its resources than attempting to assist those who had been in Canada for some years, were familiar with the ways of the country and who had the wherewithal to purchase land from earlier established English and French settlers, as occurred in some of the smaller settlements in Manitoba (JCA Fond 10032; KC, MA-1-KC-146).

The JCA could have been a major force in the colonization of Western Canada had it not geographically dispersed its financial resources so widely, attempting to aid too many projects both on a global scale and regionally within Canada. This diffusion of effort was exacerbated by Baron Maurice de Hirsch's policy of direct oversight from Paris, his views on the nature of aid and the reactive stance of the JCA administration. Aid was seldom sufficiently focused nor was it sufficiently generous to give under-capitalized settlers the boost they needed to weather the vagaries of prairie agriculture.

Other immigrant groups that successfully achieved nucleated settlement received special consideration from the Dominion Government by way of reserves of land set aside for their exclusive use. The Mennonites were granted such reserves in Manitoba and Saskatchewan; the Doukhobors had reserves in Saskatchewan, which permitted them to achieve nucleated settlement and contiguity of land holdings. In southern Alberta, the Church of Jesus Christ of Latter-day Saints (Mormons) partnered with various corporations to obtain large blocks of land where every quarter was made available for settlement (Katz and Lehr 1991). Without the government retaining land for disposition under the homestead system, Mormons were freed from the usual residency requirements and could settle in large farm-villages, thereby attaining socio-religious objectives denied them under the free homestead system and attaining the population thresholds that allowed their social and religious institutions to flourish. The Jews lacked a corporate champion with financial clout or an individual with the moral authority of Leo Tolstoy, who advocated for the Doukhobor cause. The Jews had no effective Canadian advocate and had to contend with underlying suspicion from the Christian political establishment. Without either, there was never a realistic hope that the Dominion Government would grant a reserve of land for the exclusive settlement of Jews.

While the JCA's decision to have Jewish settlers work within the framework of the Dominion Lands Act to secure land was fiscally prudent, enabling it to retain its capital to render aid to a wider array of settlers, it was socially unfortunate. Securing territory where the limitations of the Dominion Lands Act did not constrain settlement behaviour would have ameliorated the conflict between religious observances and dispersed agricultural settlement by enabling attainment of a critical mass, allowing the development of mature Jewish institutions that met the needs of observant settlers.

A major shortcoming of the JCA was its failure to think strategically. Because its mandate was to be a philanthropic agency for Jewish refugees, it dispersed its funds to respond to short-term crises. In Canada, it had no strategic plan. Although it advanced more than 2,500 loans to Jewish farmers in Canada from 1907 to 1930, without a clear focus, it pursued a scattergun approach that helped many a little rather than a few adequately (Belkin 1931). Most JCA aid was in the region of two or three hundred dollars, sufficient to buy a cow or perhaps a yoke of oxen but little else, and this aid further burdened the settler with interest payments. Capital was in short supply on the agricultural frontier. It was hard to generate without off-farm work because markets able to absorb increased production were not always readily accessible. The JCA never seemed to truly appreciate the harsh reality of frontier economics and the difficulty of transitioning from a subsistence to a market economy. It also failed to grasp the importance of creating compact settlements that could achieve the critical mass necessary for Jewish religious life to be self-sustaining.

There is a caveat to all this: the Jewish faith does not have a rigid hierarchical structure as do most Christian denominations. Unlike most other religious groups, such as the Mormons, whose society is theocratic, Judaism has no central administrative council; authority devolves to a multitude of rabbis, each of whom is independent. Friedgut (2009, pp. 41–2) points to the lack of rabbinical authority contributing to community dysfunction in Lipton as being one of the many factors contributing to the colony's decline. It is one of history's great ironies that, for lack of a theocratic structure, Jewish settlement more easily fell victim to the demands of theological principles. Without a theocratic hierarchy, Jews were disadvantaged in their negotiations with governments and were reliant upon secular institutions such as the distant JCA to advance their cause. Ironically, the seeds of the decline of Jewish agricultural settlement in Western Canada derived from the demands of their religion.

While there is a plethora of reasons why Jewish agricultural settlements ultimately atrophied, among which the structure of Jewish society and the demands of its religion, locational factors should not be ignored. Because the JCA was not active in Western Canada until 1907, it bore no responsibility for the locations chosen for settlement by Jewish settlers. This saw Jewish settlements scattered across the prairies with colonies such as Lipton, Sonnenfeld, Hirsch, Rumsey and Wapella, all located within Palliser's triangle, the sub-humid part of the prairies most subject to drought and where mixed farming was most difficult (Figure 4.1).

The story of JCA involvement in Western Canada was not one of total failure. It did what it could within the limits imposed by the philosophy of its founder, the religious and cultural aspirations of its beneficiaries, the restrictions of Canadian government policy and the unfortunate positioning of many settlements within the drought-prone Palliser's triangle.

Nevertheless, its story is one of missed opportunities by a well-intentioned settlement organization that was too distant to be efficient and gave too little, too late, to be effective.

References

Arnold, A J 1968, "The contribution of the Jews to the opening and development of the West", *Transactions of the Historical and Scientific Society of Manitoba*, Series 3, No. 25, pp. 23–37.

Arnold, A J 1972, "The life and times of Jewish pioneers in Western Canada", *A Selection of Papers, The Jewish Historical Association of Western Canada*, Second Annual publication, Winnipeg, pp. 51–7.

Avni, H 1991, *Argentina and the Jews: A History of Jewish Immigration*, University of Alabama Press, Tuscaloosa.

Belkin, S 1931, "The Jewish Colonization Association: forty years of immigration, land settlement and education", *Canadian Jewish Chronicle*, December 11.

Belkin, S 1966, *Through Narrow Gates: A Review of Jewish Immigration, Colonization and Immigrant Aid Work in Canada (1840–1940)*, Canadian Jewish Congress and the Jewish Colonization Association.

Feldman, A 1982, "Sonnenfeld – Elements of survival and success of a Jewish farming community on the prairies 1905–1939", *Canadian Jewish Historical Society Journal*, Vol. 6, Winter, pp. 33–53.

Feldman, A 2003, "Were Jewish farmers failures? The case of township 2-14 W 2nd", *Saskatchewan History*, Vol. 55, No. 1, pp. 21–30.

Fogarty, R 1975, "American communes 1861–1914", *Journal of American Studies*, Vol. 9, No. 2, pp. 145–62.

Friedgut, T 2007, "Jewish pioneers on Canada's prairies: the Lipton Jewish agricultural colony", *Jewish History*, Vol. 21, pp. 385–411.

Friedgut, T 2010, "The Lipton Jewish agricultural colony 1901–1951: pioneering on Canada's prairies", *The Inaugural Lecture of the Switzer-Cooperstock Lecture Series*, Jewish Heritage Centre of Western Canada (JHCWC), Winnipeg.

Gitelman, Z 1988, *A Century of Ambivalence: The Jews of Russia 1881 to the Present*, Schocken, New York.

Gutkin, H 1980, *Journey into Our Heritage: The Story of the Jewish people in the Canadian West*, Lester and Orpen Dennys, Toronto.

Israelite Press 19 June 1923, JHCWC.

Jewish Times 6 July 1900, JHCWC.

Katz, Y and Lehr, JC 1991, "Jewish and Mormon agricultural settlement in Western Canada: a comparative analysis", *The Canadian Geographer/Le Géographe Canadien* Vol. 35, No. 2, pp. 128–42.

Katz, Y and Lehr, JC 1993, "Jewish pioneer agricultural settlements in Western Canada", *Journal of Cultural Geography* Vol. 14, No. 1, pp. 49–67.

Lambrecht, KN 1991, *The Administration of Dominion Lands 1870–1930*, Canadian Plains Research Centre, Regina.

Lehr, J 1977, "The government and the immigrant: perspectives on Ukrainian block settlement in the Canadian West", *Canadian Ethnic Studies*, Vol. 9, No. 2, pp. 42–52.

Lehr, J 2009, "'A Jewish farmer can't be': land settlement policies and ethnic settlement in Western Canada 1870–1919", *Jewish Life and Times*, Vol. 9, pp. 18–28.

Leonoff, C 1982, *Pioneers, Ploughs and Prayers: The Jewish Farmers of Western Canada*, Jewish Historical Society of British Columbia and the Jewish Western Bulletin.

Leonoff, CE 1984, *The Jewish Farmers of Western Canada*, Jewish Historical Society of British Columbia, Vancouver.

Lousada, HG 27 March 1906, "HG Lousada, London, to Lord Strathcona, High Commissioner, London", Library and Archives of Canada, Ottawa (LAC), RG. 25 A-2 Vol. 142 C2/91.

Martin, C 1973, *"Dominion Lands" Policy*, LH Thomas (Ed), McClelland and Stewart, Toronto.

Norman, T 1985, *An Outstretched Arm: A History of the Jewish Colonization Association*, Routledge and Kegan Paul, London.

Oleskiv, I 1895, *O Emigratsii* [On Emigration], Oshchestvo Mykhaila Kachkovskoho, Lviv.

Richtik, J and Hutch, D 1977, "When Jewish settlers farmed in Manitoba's Interlake area," *Canadian Geographical Journal*, Vol. 95, No. 1, pp. 32–5.

Robinson, L 1912, "Agricultural activities of the Jews in America", *American Jewish Yearbook*, Vol. 14, pp. 21–115.

Rosenberg, L 1939, *Canada's Jews: A Social and Economic Study of the Jews in Canada*, Bureau of Social and Economic Research, Canadian Jewish Congress.

Rosenberg, S 1970, *The Jewish Communities in Canada*, Vol. I, McClelland and Stuart, Toronto.

Sack, B 1965, *History of the Jews in Canada*, Ralph Novek (trans.), Harvest House, Montreal.

Sanders, A 1988, "The history of the Jewish farming colony at Rumsey, Alberta", unpublished typewritten manuscript ca. 1968, JHCWC, File 154 F9.

Schwartz, E and Te Velde, J 1939, "Jewish agricultural settlement in Argentina: the ICA experiment", *The Hispanic American Historical Review*, Vol. 19, No. 2, pp. 185–203.

Shpall, L 1950, "Jewish agricultural colonies in the United States", *Agricultural History*, Vol. 14, pp. 120–46.

Spry, I 1959, "Captain John Palliser and the exploration of Western Canada," *The Geographical Journal*, Vol. 125, No. 2, pp. 149–84.

Stein, A 2001, "The Jewish farm colonies of Alberta", *Canadian Jewish Outlook*, March/April, p. 39.

Tennenhouse, L 1972, Transcript of recorded interview, March 10, JHCWC, Tape No. 217.

Tulchinsky, G 1992, *Taking Root: The Origins of the Canadian Jewish Community*, University Press of New England, Hanover and London.

Zablotsky, E 2004, "Philanthropy vs. unproductive charity: The case of Baron Maurice de Hirsch", *Serie Documentos de Trabajo*, Universidad de CEMA, No. 264.

Archival sources

AM – Archives of Manitoba, Winnipeg, Township General Registers; Records of Homestead Entry

CJA – Canadian Jewish Archives, Montreal, JCA Fonds

JHCWC – Jewish Historical Centre of Western Canada, Winnipeg

LAC – Library and Archives of Canada, Ottawa

5 Soldier settlement and the expansion and contraction of the bush frontier in New Zealand during the 1920s

Michael Roche

Introduction

The 1920s witnessed the end of a phase of pioneering land settlement in New Zealand that had extended beyond the point at which it could be sustained and after which there was retreat from some areas of the hill country of the North Island in the face of scrub regrowth and accelerated soil erosion. The Liberal Government (1891–1912) had initiated the expansion of pastoral farming into the forested hill country of the North Island as part of a closer land settlement policy. This work was further carried forward by the Discharged Soldier Settlement Act of 1915, which is the subject of this chapter.

In excess of 110,000 men out of a total population of 1.1 million served in the New Zealand forces during World War I, with 18,000 dead. Conscious of a need to meet a debt of honour, the government passed the Discharged Soldier Settlement Act in 1915. By the mid-1920s, the "margins of settlement" had been, in places, exceeded and the frontier was rolled back. In New Zealand, it was not drought and wheat, as in South Australia, but rainfall and pasture deterioration that effected what Cumberland (1941, p. 554) termed "nature's revenge".

This chapter presents some macro-level data that chart the ebb and flow of the pastoral frontier between the wars before outlining the soldier settlement scheme and the land problem facing the government from 1917 into the 1920s. Attention is then turned to the regional level and to Wellington Land District, where soldier settlement was pushing into the forest frontier. There follows a case study of a bush frontier soldier settler from the Upper Whanganui River District. The chapter concludes with an assessment of soldier settlement on this bush frontier.

Ebb and flow of the settlement frontier

New Zealand was approximately 53% forest-covered in 1840. After 60 years of European settlement, this figure was reduced to around 25%, with most of the clearance taking place in the last 15 years of the nineteenth century,

largely coinciding with the election of a Liberal government and implementation of a large-scale closer land settlement policy that saw pastoral farming push up into the heavily forested hill country of the North
Island. At a national level, this expansion and retreat is captured by statistics of total occupied lands, moving from 28,169,788 acres in 1886 to
a peak of 43,546,757 acres in 1921 and then declining to 42,901,772 acres
in 1939 (*Statistics of New Zealand* 1911; *New Zealand Official Year Book*
1924, 1941). This expansion and withdrawal is more dramatic than it at first
appears because it was concentrated in the western central North Island.
The retreat of the pastoral frontier can be captured at the regional level
through statistics on the expansion of fern scrub and secondary growth
on unimproved occupied land. In Wellington Land District in 1921, these
amounted to 309,999 acres; by 1927, the figure was 330,775 acres; by 1930, it
was 359,316 acres and 428,034 acres in 1940 – or an increase from 23.6% of
the total unimproved occupied land in the district in 1921 to 31.4% by 1940
(*New Zealand Official Year Book* 1924, 1941). State responses to this roll-
back of settlement were piecemeal until 1925 when a Deteriorated Lands
Act was passed and a Special Committee of Inquiry set up to report on the
problems facing farmers, particularly in the West-Taupo and Whanganui
hill country. This report noted that 11,000,000 acres of hill country pasture
comprised 75% of the total deforested lands in New Zealand. Of this area,
"during the last eight years, according to official statistics, approximately
1,100,000 additional acres have reverted to fern, scrub and secondary
growth" (AJHR 1925, C15, p. 10).

Discharged soldier settlement

In 1915, William "Farmer Bill" Massey, the prime minister, introduced
a Discharged Soldier Settlement Bill into Parliament. Once enacted into
law, this enabled discharged members of the New Zealand Expeditionary
Force or naval forces who had served overseas to ballot, to lease or purchase
land[1]. Government loans were available to establish the farm and build a
house. Nearly all of the discharged soldiers opted to lease rather than buy
the properties they selected. From the first, the Commissioners of Crown
Lands (CCL), who were the district heads of the Lands Department, recognized that the regulations "practically make it impossible for the Lands
Board to debar soldiers without farming experience" (AJHR 1917, C9, p. 9).
The scheme was closed in 1941, long after the years of peak activity in 1919
to 1922 when, of some 12,694 applications[2], 2,270 (or 18%) were granted for
a total of 1,033,748 acres (Table 5.1).

From its inception in 1915 to 1933, Williams (1936, p. 126) noted that
there had been 4,071 allotments totaling 1,432,690 acres but that sales forfeiture and abandonment had reduced these to 2,727 holdings on 943,551
acres. These figures highlight, rather incidentally, the comparative size

Table 5.1 Applications and allocations under the Discharged Soldier Settlement Act, 1915–1941

Year	Applications Received	Number of Allocations	Area of Allocations (Acres)
1916	272	2	629
1917	522	319	143,524
1918	513	313	103,362
1919	1379	348	117,018
1920	5041	932	403,891
1921	5396	1087	414,867
1922	878	403	97,972
1923	284	146	25,113
1924	216	79	16,910
1925	1123	47	9014
1926	109	86	20,500
1927	78	66	17,412
1928	96	60	15,695
1929	90	77	13,275
1930	63	53	16,665
1931	41	31	8495
1932	22	16	4125
1933	12	6	4133
1934	10	8	1536
1935	8	7	2954
1936	12	7	2638
1937	8	8	1877
1938	1	4	315
1939	1	1	11
1940	2	2	323
1941	2	2	220
Totals	15,179	4,110	1,443,564

Source: AJHR, 1941, C9, p. 2

of the pulse of land development under the scheme in the period 1919 to 1922 (2,770 allocations of 935,777 acres) and the subsequent falling back in occupied area over the next 18 years. Williams' data indicated that 66% of the area allocated and 67% of the number of settlers who had entered the scheme had departed within 18 years, and this reinforced other period views of the scheme as failing to meet the debt of honour. The scheme can be studied as part of the wider historical geography of land settlement into which it is subsumed according to Williams (1936, p. 126), who observed that, by the mid-1930s, soldier settlement in New Zealand was "an 'Historical episode' and there no longer appears to be any valid reason for a separate scheme" outside of the Lands for Settlements Act. Arguably, the scheme's importance to the ebb and flow of the pastoral frontier meant that they still deserved separate treatment in the 1920s – and also from present-day researchers.

Planning and management of the Soldier Settlement Scheme

Massey realized early on that there was insufficient available Crown Land to meet the needs of returning soldier settlers. In 1919, these comprised 704,178 acres already open for selection to other Crown settlers along with another 3,760,356 acres available across the entire country (*New Zealand Official Year Book* 1919, p. 487). The national mean-sized holding in 1920 was 532 acres, so that the available land, in terms of averages, might represent only just in excess of 7,000 farms. Against this, there were some 90,000 demobilized men and, even if half returned to their previous or urban occupations, 3.7 million acres was sufficient for just 15% of the remainder. Additional impetus came from Massey's vision of New Zealand as a rural society of independent small farmers[3]. More land was needed, and here the government could fall back on the practices of the earlier Liberal Government and repurchase areas of freehold land for lease or sale back to soldiers. The seeds of the future problems were laid in these initial conditions of there being too little Crown Land available for settlement and of its being, by its very nature, remote and of poorer quality. Other repurchased land was sometimes expensive and also not always of good quality for pastoral farming. The soldier settlement legislation was amended in 1917 to permit discharged soldiers to borrow from the government to purchase existing farms known colloquially as "Section 2" farms. This measure was intended to take the pressure off Lands officials struggling to throw open Crown Lands at the bush frontier and to enable others who wanted farms to be absorbed more easily and quickly into the existing farming landscape[4].

A balloting system previously used in the closer land settlement era was carried across to the soldier settlement allocations. While many claimed to be farmers, they were more likely to have been farm labourers. Some unsuitable men thus entered the system. In an era when the traumatic stresses of frontline service were not understood, others who had suffered from "shell shock" also turned to farming with predictable results.

Once on the land, the soldier settlers were monitored fairly regularly by the Crown Lands Rangers, who offered advice about stock purchase as well as noting improvements to the properties, such as fencing, clearing bush and constructing a house and farm buildings. In addition, the regional CCL solicited their opinions in making decisions on approving or declining loans to soldier farmers. As economic conditions deteriorated in the early 1920s, the State assisted soldiers in staying on the land by offering a range of rent adjustments and postponements, debt consolidation, a revaluation of the capital value of the property as well as writing off some debts. Against the deeply felt popular rhetoric, which endures to this day, about men "walking off" the land and being regarded as quitters in the popular imagination, State efforts to keep settlers in place were greater than was recognized at the time and most certainly are today. An unintended result

was that some soldier farmers and their families survived on their properties but had virtually no equity in the farm after 20 years of occupancy. The
State could also save the settlements by amalgamating farms to form larger
economic units in order for them to survive deteriorating primary produce
export prices. This strategy was employed on stock and crop, rather than
dairy, farms. It could ensure that a contiguous farm settlement survived
but that not all of the settlers could remain and some departed with little
more than no debt.

Changing fortunes for discharged soldier settlers in the 1920s

The "failures" of the scheme have been thought about differently by officials
and politicians over time. Williams (1936), as alluded to previously, subsumed soldier settlement within the broader category of Crown settlers by
the mid-1930s. This was a defensible position on the grounds that most of the
farms had been taken up 15 to 20 years previously and because the financial
problems the settlers faced were, by then, a result of the Great Depression
rather than of the ex-soldier status of the farmers. It was less valid in the
1920s when the fortunes of soldier settlement rested on a combination of
settler attributes, the environmental/economic potential and limits of the
lands allocated and the impact of the end of the Commandeer, the World
War I bulk-purchasing agreement of the British Imperial government for
New Zealand wool, meat and dairy produce that continued after the immediate end of the hostilities, finally ceasing in 1921–1923 and causing New
Zealand export prices to collapse. Ideas of failure, from an official point of
view, subtly changed as the 1920s unfolded from a focus on the success or
failure of the individual soldier farmer to the success or failure of the farm
settlements. "Persistence", the years that the original settlers spent on their
properties, is too simplistic a yardstick that does not allow for an economically sensible behaviour pattern, such as leaving the farm before incurring
yet more debt (for instance, on dairy farms where financial authority to borrow for stock purchase was so delayed that the milking season was over,
and there would be no income for a further year) or for capturing occasions
where families stayed on the land for many years in penury.

Land Revaluation

By 1922, a sharp decline in primary export prices meant that many soldier
settlers could not pay their rents or the interest owing on other government
loans. One State response was to establish a Dominion Revaluation Board
to which soldier settlers could apply for a new capital valuation of their
land. A new lower valuation reflecting current market conditions meant a
reduction in rents and, along with other measures such as remission in payments for several years and other debt restructuring, was designed to keep
the soldier settlers on their farms. Nationally, 5,284 cases were dealt with by
the Dominion Revaluation Board, and the capital value of leasehold soldier

farms was reduced by a total of £1,615,160 (AHJR 1925, C9, p. 4). The CCL for Wellington commented that, "The land-values are [now] more in keeping with present-day ruling prices" (AJHR 1925, C9, p. 16).

For Crown settlers of all types in the Upper Whanganui and West-Taupo areas, legislation in 1925, in the form of the Deteriorated Lands Act, provided further financial relief. This act recognized that pioneer settlement in the heavily forested hill country was failing through degeneration of the newly established pasture by regrowth of indigenous ferns and scrub and by accelerated soil erosion in the wake of the removal of the original forest cover. Under this legislation, Crown selectors, including soldier settlers, could apply for a revaluation of the land and other debt restructuring concessions. The districts under review amounted to 874,700 acres including 1990 farms, largely on leasehold tenures on Crown Land. Abandoned farms numbered 75, totaling 42,905 acres or 5% by area. The area of reverted land was estimated at 232,500 acres or 27% of the total (AJHR 1925, C15). The necessity for revaluation of the land rested on its being prone to secondary regrowth, the costs of using sufficient numbers of cattle (they grazed widely, not in concentrated patches like sheep and ate and trampled the regrowth to maintain pasture), the lower price of poorer quality bush wool and the costs of lack of access (AJHR 1925, C15, p. 3). The records of farm inspections carried out under the Dominion Revaluation Committee and Deteriorated Lands Act provide rare "snap shots" of individual properties at two points in the 1920s[5].

Flowing on from the activities of the Dominion Revaluation Board, a change of focus from "farmers" to "farms" is evident in the government annual reports on soldier settlement that, from 1917 to 1924, contain considerable detail about transfers of land but that, after 1925, have rather more on mortgages and revaluation than on sums borrowed and repaid.

Typologies of failure

Three temporal episodes of failure of the scheme have been identified of which two are pertinent here. The first was that of the "failure of the soldier settler", which was prevalent in official and popular circles in the 1917 to 1922 period and contrasts with the "failure of the farms" period of 1923 to 1928. The "failure of the settlers" attributed farming failure to the character, experience and lack of effort by the soldier settlers, particularly where leases were forfeited. The "failure of the farms" was a belated acknowledgment that some properties were too small to be economically farmed under more difficult economic conditions and/or that the land was not well suited for farming. Thereafter came a structural failure phase (1929 to 1939) caused by the impact of the Great Depression (Roche 2008). This typology was developed from a reading of existing general surveys and more detailed archival research on two regions, Canterbury and Wellington. As such, it provides some guidance for thinking about the case study that follows. But the typology was not specifically oriented towards the soldier settlers on the bush frontier because they have erroneously become much too emblematic of the scheme (e.g., Cumberland 1981). Politicians also used military imagery to depict soldier settlers as battle-hardened warriors now taking over and transforming unproductive forestland into valuable farmlands. In this chapter, attention is directed back at the bush frontier where soldier settlement was the advance party of continued land settlement and was central to the enduring popular images of the scheme.

Wellington Land District – soldier settlement on the frontier

At a regional level, soldier settlement and the bush frontier can be tracked statistically by examining the numbers and acreage of Ordinary Crown Land taken up under the Discharged Soldier Settlement Act. This brings to light one of the difficulties of summarizing the extent of the scheme because ex-soldiers under the act could be placed on land of many differing tenures. Those who selected farms on Ordinary Crown Land were more likely to be at the forest frontier, for these lands had previously been Maori land. By contrast, the Settlement Lands had been previously alienated but were offered back to the State for reallocation and thus were more likely to be within the margins of settlement, albeit that some might still be remote and undeveloped. Wellington Land District stretched from Wellington City northwards to the west of Lake Taupo (Figure 5.1). It contained a sizable amount of remote bushland, part of the core of the bush frontier that extended across parts of the Taranaki, Auckland and Wellington Land Districts, thus posing further data difficulties when discussing pioneer land settlement at the forest margins of the central North Island, which spanned portions of several land districts.

Considering Wellington Land District in more detail (Table 5.2), it can be seen that soldier settlement on Crown Lands commenced at a sizable

Table 5.2 Lands allocated under the Discharged Soldier Settlement Act in Wellington Land District (YE 31 March)

Year	Ordinary Crown Lands		Settlement Lands		National Endowment Lands	
	No.	*Acres*	*No.*	*Acres*	*No.*	*Acres*
1917	64	23601	60	10467	-	-
1918	30	15860	60	9987	-	-
1919	20	12700	58	9452	-	-
1920	53	22171	174	17403	-	-
1921	29	5383	166	45572	1	878
1922	30	15104	84	12960	-	-
1923	12	1774	15	1224	-	-
1924	6	1014	16	2374	-	-

Note: National Endowment land set aside to provide rents to fund education and pensions were found largely in the South Island.

Source: AJHR, C9, 1917–1924

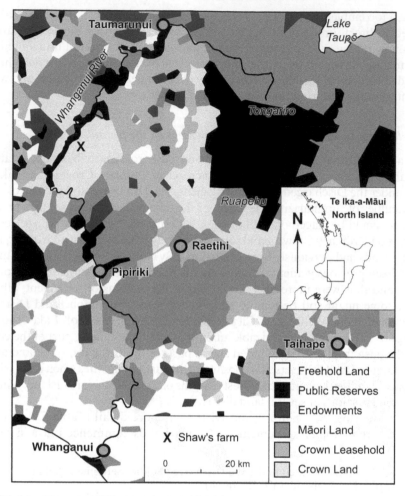

Figure 5.1 Land tenure Upper Whanganui Region.

Source: AJHR C1 1916

scale in terms of area as early as 1917, ahead of the main pulse of settlement from 1919 to 1922. The latter pulse is also revealed as being accommodated mainly on Settlement Land rather than Crown Land, and that this was formerly freehold land offered to and repurchased by the Crown for redistribution as soldier settlements. These Settlement Lands had already been farmed and were generally not on the forest margins.

In 1918, the CCL for Wellington reported on 22 soldier settlements, of which five were at the forest frontier of the Upper Whanganui and West-Taupo regions (Table 5.3). The exception was Puektoi Settlement in the Puketoi Ranges in the northern Wairarapa to the south, which was not part of the main forest area of the central North Island. In the case of the

Table 5.3 Bush Frontier Settlements under the Discharged Soldier Settlement Act in Wellington Land District to 1919

Settlement	Acres	No. of Farm Sections	Vegetation	Felled	%	*Grassed	%	Cattle	Dairy Cows	Sheep	Pigs	Horses	People
Owhango Piriaka	2730	16	Milled out forest	2009	74	1785	65	456	132	1065	10	10	43
Manganui	606	2	Bush	246	41	254	42	45				3	2
South Waimarino	40,775	48 (31 selected)	Virgin bush	1200	3	423	1	13	1	650		2	?
Kariori	598	2	Bush	5	0.8	210	35	37	75		5		5
Puketoi	506	3 selected	Bush	130	26	20	4						

*Grassed is presumably a subset of the felled area, whereas for the other settlements these are recorded separately.

Source: AJHR, C9A, 1919, 5

Owhango and Piriaka settlements, these were former forest reserves that had been milled and were now turned over to land settlement; the others were at the forest frontier where the land settlement potential was considered to outweigh the value of the heavy forest cover. In 1918, the conversion of forest to pasture by felling and burning was just under way and limited numbers of stock were on the properties. The small number of human occupants points to the comparative isolation of these bush frontier settlements. Progress was briefly reported on by the CCL in the annual reports on soldier settlement from 1920 to 1922. In 1920, it was recorded that approximately 100 settlers had taken up bush land around Raetihi, Owhango, Piriaka and in the Retaruke and Whirinaki districts with more than 10,025 acres being sown in grass. Two partially failed burn offs were noted, but the CCL for Wellington's tone was exceedingly optimistic. "There is little doubt that in the course of few years these settlers will be in a most enviable position" (AJHR 1920, C9, p. 10). The next year the report contained some similar sentiments: "16,363 acres in grass, and other improvements, such as fencing and buildings making a total value of £98,710" (AJHR 1921, C9, p. 12). But there was also the concerning note that "the prospects of these settlers were undoubtedly bright, but owing to the extreme drop in the value of sheep and cattle and the price of wool, they will not in the near future obtain the returns that were expected" (AJHR 1921, C9, p. 12). This was followed in 1922 by the observation that the year had been "most disastrous for all those engaged in pastoral and dairying pursuits" (AJHR 1922, C9, p. 13).

In the absence of a suitable map showing vegetation cover in the 1920s for the Upper Whanganui and northern portions of the Wellington Land District, a land tenure map serves as a proxy (Figure 5.1). The map shows the extent of freehold land successively added to since European colonization in 1840 and a core of Maori land in the central North Island that had been divided by a zone of land purchases by the Crown. Virtually all of this area was forested in 1840 apart from some tall tussock grassland to the north and east of Mt. Ruapehu and Mt. Tongariro, a consequence of earlier volcanic eruptions that had destroyed the forest cover. Maori land and Crown Land shown on this 1916 map were largely under heavy forest[6]. By 1916, what was originally a large contiguous area of Maori land in the central north Island had been split into two, separated by an expanding enclave of Crown Land. For most of this period, only the Crown could purchase Maori land. The block freehold lands to the south (predominantly European owned and deforested from the 1860s), within which more intensive pastoral farming was carried out, abutted the southerly area of Maori land, which was surrounded to the north by a mixed zone of Crown Land, Crown leasehold and other reserved Crown Lands mostly in forest. Some of the Crown Land was judged to be suitable for settlement and was taken up as leasehold farms; other land was held in reserve as Crown Land or deemed unsuitable for settlement. Still other land was forest reserve accessible to sawmillers or set aside for felling in the future. A large area shown in Figure 5.1 as public reserves was also set aside as Tongariro

National Park. What the map shows quite effectively is that pioneering soldier settlement on Crown Land at the bush frontier took the form of a series of settlement enclaves rather than a single line of settlement rolling northward towards the remaining large area of Maori land.

To explore the soldier bush farming frontier further, a case study of a single farm in the Upper Whanganui region follows (space precludes the inclusion of others).

Jim Shaw, soldier settler

William James Henry Mitchell (Jim) Shaw (1876–1938) spent his early years in rural Otago and Canterbury employed in varied occupations from farm labourer to brick maker before becoming a hotel keeper in 1911. On enlistment at age 40, in May 1916, he was married with three teenage children and gave his occupation as "miner." Shaw initially volunteered for the New Zealand Tunnelling Company but later served as a sapper before being transferred to the New Zealand Machine Gun Company on the Western Front (Figure 5.2). In April 1918, he was discharged as unfit for further military service on account of an old leg fracture and chronic rheumatism that made it impossible for him to march (Shaw, AABK 18805 W5553 13/0103785).

Early in 1919, Shaw was working for a Dr. Wall on a dairy farm near Foxton in the Manawatū but later in the year successfully balloted for

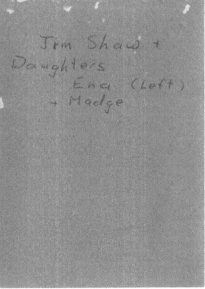

Figure 5.2 Jim Shaw in uniform with his daughters Ena and Madge, 1916.

Source: Wairarapa Museum Collection, George Hardy Photo Album, 15-154/160

1,419 acres comprising Section 3, Block XIV, Retaruke Survey District, on the Whanganui River. It was a fragment of a large 458,000-acre plot of Maori Land purchased in 1887 (*New Zealand Gazette* 1887, pp. 677–8). Shaw's property bordered the Mangapurua Valley soldier settlement of some 40 farms. By 1940, these had all been abandoned, some unwillingly (Bates 1982). An iconic "bridge to nowhere" now stands there amongst regenerating bush and is itself a tourist attraction[7].

Shaw, on his application, claimed seven years' experience on Glenmore Station in the South Island, bush-clearing work and other general farming experience, but he had only £50 in cash. That year, in Wellington Land District, 79 applicants had no money and 46 had £50 or less, but 96 possessed £100 to £500 (AJHR 1919, C9, p. 4). The official view was that those with little capital stood the best chance of success on "forest covered Crown Land of good quality or first class dairy land" (AJHR 1917, C9, p. 9). These circumstances describe the conditions faced by Shaw.

Shaw had taken up his land sight unseen, relying on the recommendation of a friend already established on a neighbouring property. The entire district was heavily forested. Section 3 was very isolated, effectively only possessing river access. One of Shaw's earlier actions was to co-lease another four acres by the Whanganui River as a landing site and holding paddock for stock (Rutherfurd to CCL 1920). The initial Crown Lands Ranger's reports were positive. Shaw was described as occupying "a heavy bush section, but [he] is making good progress and will so be able to stock the area already sown [in grass]" (CCL to Under Secretary for Lands, 1921). A loan for £600 had been approved the previous year for bush felling and grass seed purchase (Table 5.4). Over the next two years, Shaw borrowed another £650, bringing him to the £1,250 maximum limit. By the end of 1921, he still had only debts from his efforts. In 1923 and 1926, he unsuccessfully approached the department for additional sums to purchase stock but was declined on the grounds that the value of his improvements (fencing, area in grass and a house) were insufficient to warrant further loans.

Table 5.4 Shaw's loans and expenditure, 1920 to 1922

	Loans		Expenditure					
Approval of Loan	*Individual Loan £*	*Cumulative Loan £*	*Bush Felling £*	*%*	*Grass Seed £*	*%*	*Improvements £*	*%*
17/12/1920	600	600	400	66.6	200	33.3	-	-
21/10/1921	400	1000	-	-	-	-	400	100
14/12/1921	200	1200	-	-	200	100	-	-
6/2/1922	50	1250	-	-	-	-	50	100
Total by type			400	32	400	32	450	36

Source: Compiled from AAMA Acc W4320 36 RL203 WJHShaw Sec 3 Blk XIV Retaruke

Officials laid some of the blame at Shaw's feet. Early in 1922, the Crown Lands Ranger noted that Shaw's 21-year-old son James was actually managing the farm, with his father then, unusually in terms of the scheme, still running Dr. Wall's dairy farm. The ranger was scathing about James' handling of an expensive bush felling contract; "had he any sense he would have cancelled the contract; but being the most useless and lazy waster in the block he let things drift" (Rutherfurd to CCL 1922). Shaw's son had been on the farm since 1920 but, in the ranger's opinion, had done little apart from building a whare[8] and packing in grass seed, "He has done no work not even setting out in a garden. The section requires stocking" (Rutherfurd to CCL 1922). Stock and station agents[9] Wright Stephenson and Co., in pressing the Lands Department for settlement of other debts incurred by Shaw, claimed that he had "one of the best sections on the place", and that they were aware of other settlers who, with poorer land and smaller farms, had obtained extra advances beyond the £1,250 (Wright Stephenson and Co. to CCL 1922). Shaw faced a dilemma in that he needed stock but was unable to raise the capital to purchase them. In 1923 and now in residence, Shaw remonstrated to the officials that he, his wife and two daughters had worked hard "to get ahead". He had put on 40 head of cattle, but his pasture had deteriorated and it was imperative to have sheep; bitterly he wrote that "if you had got me some help when I asked you there would have been money off the place by now" (Shaw to CCL 1923). By 1924, Shaw had sent his family away and remained alone on the farm. He succeeded in stocking it with 450 breeding ewes, 200 mixed lambs and 7 rams but under lien to his old employer, Dr. Wall. That is, Wall had loaned him money, but the sheep that Shaw purchased were the security on the loan so, on selling them at ruling prices and repaying his debt, he was no further ahead. His receipts in 1924 came to £864 against an outlay of £1,041 (Craig to Under Secretary of Lands 1924).

The patience of officials was now becoming stretched – "in the past Mr Shaw's methods of farming has not impressed me" wrote the Crown Lands Ranger (Rutherfurd to CCL 1927). In particular, the ranger pointed to Shaw's son James' inactivity and to the fact that Jim Shaw, for the past 15 months, had been employed by the Public Works Department in road-making duties. The section, the ranger declared, was "overrun with wild pigs and it will now take a better man than Shaw to cope with them" (Rutherfurd to CCL 1927). The 1927 report of the Deteriorated Lands Committee on the property noted that Shaw was "not a very satisfactory settler" (CCL to Special Valuation Committee 1927). In 1927, Shaw had 220 sheep on the property and 20 cattle but none in 1928. As forfeiture of the lease moved ever closer, the CCL described the situation as one where "there is no hope of Shaw doing any good" (CCL to Under Secretary of Lands 1928).

It is, however, possible to reread Shaw's story from other than the officials' viewpoint. In continuing to work on Wall's farm as long as possible, he was trying to bring more capital to the farm. He did seem to overestimate his son's capacity, but this makes no allowance for James' youthful inexperience and the extreme isolation of the property. Shaw was alert to the need to stock the farm but, having reached his loan limit, was never able to make enough improvements on the farm to borrow more – hence his employment with public works to generate family income.

The overly promising initial appraisals of Shaw's land and its dramatic reappraisal are trackable through the spectacular decline in capital value of the section by the Dominion Revaluation Board from £2,130 in 1919 to £1,760 in 1924, £1,540 in 1926, £540 later that same year and finally to £354 by the Deteriorated Lands Committee in 1927. In this series of downward revaluations, there are additional clues as to the environmental limitations of the property – unrecognized by officials in 1919. The ranger's inspection in 1923 on the one hand noted Shaw had "been doing good work latterly" but also that "the place will soon go back and is very bad for silver fern" (Memo to Land Board 1923)[10]. The inspection of the property after Shaw applied for relief under the Deteriorated Lands Act was telling. The pasture was described as being badly deteriorated with much second growth. The report noted the section was "very steep and broken country sandstone formation cleared area rapidly reverting to second growth it is only a matter of a few years & the improvements will be valueless" (Hall to CCL 1927). By 1928, 150 acres of felled forest and 180 acres previously sown in grass had reverted to secondary growth, while 1,088 acres remained in standing bush – and was not worth felling (Land Board, Memo 1928).

After forfeiting the lease, Shaw continued working for a decade as a labourer for the Public Works Department, on road maintenance in the nearby Raetihi district, until he returned to Canterbury in 1938, where he remained as a labourer until his death later that same year. The State pursued him for repayment of rental arrears of £23/5/6d until 1929. Shaw's reply that "I have no money I lost all I had on the place besides time & have now to start life again from scratch" (Shaw to CCL 1929) was accepted and moves were made to write off the sums outstanding. The forfeited section was advertised in the *New Zealand Gazette* (1930, p. 1707) as "1,149 acres of third class land including 332 acres of felled and grassed with the remainder in rimu, matai, rata, tawhero and beech forest". Doubts were expressed as to whether anyone would ever lease the land. In 1986, the area, long reverted to bush, became incorporated into the newly formed Whanganui National Park.

The sequence of victory and defeat of the soldier settlers (and other Crown settlers) in the forested hill country of the central North Island identified by Cumberland (1941) was to be repeated by later cycles of "victory" presaged on the application of science to grasslands as well as on the beliefs that establishing a good sward of grass could counter accelerated soil erosion and, under a later national development era, that the development of aerial top dressing with superphosphate after World War II would further stabilize the slopes and increase production (Le Heron, 1989a, 1989b; Brooking and Pawson 2011; Winder 2009)[11].

Conclusion

In terms of numbers of ex-soldier farmers and the areas involved, discharged soldier settlement contributed significantly to the final phase of the expansion of the pioneer farming frontier in New Zealand during the 1920s. This took place, in the North Island in particular, on a frontier where timber on heavily forested land in higher rainfall areas had to be felled and burned before the land could be converted to pasture[12]. These forest frontier farms, concentrated initially in the central North Island, comprised a minority of soldier settlement that, more generally, took place on more benign Crown Land and Settlement Land. But they were still an important part of the last pioneering phase of bush settlement.

The soldier settlement bush farming frontier did not take the form of a Turnerian sequence and a line of continuous occupation. Instead, it formed a series of enclaves in a zone of Crown Land of a number of classes and tenures that divided and hollowed out the area between two large blocks of Maori lands that still existed in the Central North Island. This enclave pattern, if anything, accentuated the degree of isolation of some of the farms not only in a physical sense and in terms of economic linkages but also through social isolation. Added to inaccessibility was a lack of knowledge over and above that provided by many settlers having had inadequate farming experience. The old lowland deforestation techniques did not work so well in the hill country. It was especially difficult get a "hot" burn sufficient to fully clear the forest and, with insufficient stock on the land, regrowth invaded and degraded the pastures. In some instances, accelerated erosion was triggered by the clearance of forests off too steep slopes. These environmental problems were exacerbated by exogenous economic problems in the form of the collapse of wool and meat prices on the British market in 1921 to 1922. The eventual State response to this significant fall in income for soldier settlers was structural in the form of revaluations to reduce rents, postponements of arrears and interest payments, debt restructuring and, away from the bush frontier, amalgamations of adjoining properties to form more economic units. This bundle of measures also helped constitute a second zone of freehold land as a space away from the forest frontier, and often saved the farms therein, but not the farmers who might have forfeited or relinquished their leases.

The case study of Jim Shaw is a simple example of the interaction between farmer and environment, where what Cumberland termed "nature's revenge" was very quick in arriving in the form of reverted pasture. Shaw's case also illustrates the difficulties faced by farmers with inadequate capital and limited capacity to borrow other than from the State. Although he could clear some bush and sow grass, he was unable to stock the property effectively to maintain his pasture and keep secondary growth at bay. The Deteriorated

Lands Committee report also spelt out why officials felt the farming failure of men such as Shaw so acutely – "Generally speaking it may be said that strong growth of secondary scrub, &c., tells of possibilities in the country rather than impossibilities. The very factors that favour strong secondary growth alike favour good grass-growth once the sward is established, and once the secondary growth is controlled" (AJHR 1925, C9, p. 10). Although these observations about maintaining the sward anticipate the "grasslands revolution" (see Brooking and Pawson, 2011), they also harken back to Victorian ideas of progress and improvement of the natural resources of the country including the opportunity for those of limited means who were prepared to work hard to become economically independent small landholders.

Shaw had effectively "failed" before having to cope with the difficulties faced by falling export prices. His efforts to provide some off-farm income to support his family were, in retrospect, regarded by the department not as a priority but as a sign that he was poorly suited to farming. Thus, he forfeited his lease on the grounds of non-occupation of the farm in 1928. Shaw's inability to stock the farm also points to the link between environment and economics – he did not have enough capital to hold nature at bay. Finally, Shaw's experience illustrates that the two types of failure previously described by Roche (2008) are not necessarily discrete categories but overlapping ones. The official view of Shaw as a failure because he was a poor farmer overlays in time with the situation when Shaw was on a "failing farm", an uneconomic farm unit, as reflected in successive downward valuations of the land until it was eventually graded as the poorest "third class land". Such land was not easy to lease to other soldier settlers or other farmers.

Discharged solider settlement represented the end of an era of pioneering closer land settlement begun in the 1890s that, overall, saw the forest area of New Zealand reduced from around 20.3 million acres in 1893 to 12.6 million in 1923 (AJHR 1930, C3, p. 3). Although only very approximate, these figures do indicate the extent and rate of landscape transformation. What they do not show is that the expansion of pastoral farming was not without setbacks. Some significant regional frontier retreat took place in the Upper Whanganui and West Taupo districts in the 1920s as soldier settlers battled so unsuccessfully with the forest that the area of occupied lands actually shrank and the area of unimproved occupied land increased, notwithstanding the huge human efforts involved, throughout the decade.

Notes

1. Eligibility criteria were broadened over time, for instance to include those who had served only in New Zealand.
2. Applications do not exactly equal the number of farmers because some men applied in partnership with one or more for a block.
3. Even so, New Zealand was 56% urban by 1921.

4. Official sources suggest 5,403 farms on 1,227,092 acres to 1921 but, as single farms were spotted throughout the already settled farm landscape, they do not conform to typical images of soldier settlements as contiguous blocks of leasehold farms.

5. A further set of national revaluations took place under the Mortgagors and Lessees Rehabilitation Act of 1936, thus allowing the reconstruction of three synchronic cross sections in some cases. Accepting Williams' argument that, by this time, soldier settlement was subsumed within Crown settlement more generally, this latter period is not discussed further here.

6. Phillips Turner's *Botanical Survey of the Higher Waimarino District*, AJHR C11 (1909) does confirm the extent of forest cover in the district where Shaw's section was located, although the interest is in species rather than forest boundaries.

7. The section overlooked the Whanganui River rather than the Mangapurua Valley itself and, although Bates (1982) marked the farm as part of the settlement, he did not discuss Shaw.

8. Whare was the term used for a simple dwelling of possibly split-timber walls or, in this instance, more likely hand-sawn timbers.

9. Stock and station agents played a distinctive part in the Australian and New Zealand rural economies in the early twentieth century as intermediaries buying farm produce for export and importing and auctioning goods. To this were added financial services to farmers, which was particularly important if farmers were dependent on a single annual crop such as wool.

10. Silver fern, a tree fern (*Cyathea dealbata*) reaching several metres in height, is also known as punga.

11. It would be unwise to regard this as long-term "victory". There were limits to what aerial top dressing could achieve. The removal of agricultural subsidies in the mid-1980s meant an end to a long period of trying to bring as much land as possible into production, and the 200-year floods recorded in the Manawatū in 2004 provided evidence that a good sward of grass was not enough to prevent slipping of the slopes. Anecdotally, some landholders have made more money out of letting the back paddock revert to scrub and leasing space to beekeepers where the returns from premium manuka honey are better than the return from livestock farming off that same piece of land.

12. There is another last pioneering soldier settlement story to be told for the more extensive pasturelands of Otago.

References

Appendices to the Journals of the House of Representatives (AJHR), 1909, C11; 1916 C1; various years 1917–1941, C9; 1919 C9A; 1925, C15; 1930 C3.

Bates, AP 1982, *The Bridge to Nowhere: The Ill-Fated Mangapurua Settlement*, Wanganui Newspapers, Wanganui.

Brooking, T and Pawson, E (Eds) 2011, *Seeds of Empire, the Environmental Transformation of New Zealand*, IB Tauris, London.

Commissioner of Crown Lands to Special Valuation Committee under the Deteriorated Lands Act 5 October 1927, AAMA Acc W4320 36 RL203 WJHShaw Sec 3, Blk XIV, Retaruke, Archives New Zealand, Wellington.

Commissioner of Crown Lands to Under Secretary for Lands, 21 October 1921, AAMA Acc W4320 36 RL203 WJHShaw Sec 3, Blk XIV, Retaruke, Archives New Zealand, Wellington.

Commissioner of Crown Land to Under Secretary of Lands, 21 May 1928, AAMA Acc W4320 36 RL203 WJHShaw Sec 3, Blk XIV, Retaruke, Archives New Zealand, Wellington.

Craig, D to Under Secretary of Lands, Dominion Land Revalution, 24 April 1924, Commission AAMA Acc W4320 36 RL203 WJHShaw Sec 3, Blk XIV, Retaruke, Archives New Zealand, Wellington.

Cumberland, KB 1941, "A century's change: natural to cultural vegetation in New Zealand", *Geographical Review*, Vol. 31, No. 4, pp. 529–54.

Cumberland, KB 1981, *Landmarks*, Readers Digest, Surry Hills, NSW.

Hall, J to Commissioner of Crown Lands, 21 September 1927, AAMA Acc W4320 36 RL203 WJHShaw Sec 3, Blk XIV, Retaruke, Archives New Zealand, Wellington.

Land Board, Memo, Disposal of Section 3, Block XIV, Retaruke SD, 26 September 1928, AAMA Acc W4320 36 RL203 WJHShaw Sec 3, Blk XIV, Retaruke, Archives New Zealand, Wellington.

Le Heron, RB 1989a, "A political economy perspective on the expansion of New Zealand livestock farming, 1960–1984, part I – agricultural policy", *Journal of Rural Studies*, Vol. 5, No. 1, pp. 17–32.

Le Heron, RB 1989b, "A political economy perspective on the expansion of New Zealand livestock farming, 1960–1984, part II – aggregate farmer responses – evidence and policy implications", *Journal of Rural Studies*, Vol. 5, No. 1, pp. 33–43.

Memo to Land Board 18 December 1923, AAMA Acc W4320 36 RL203 WJHShaw Sec 3, Blk XIV, Retaruke, Archives New Zealand, Wellington.

New Zealand Gazette 1887, Government Printer, Wellington, No. 33, p. 677–8.

New Zealand Gazette 1930, Government Printer, Wellington, No. 37, p. 1707.

New Zealand Official Year Book 1919, Section XVII, Lands Settlement, Tenure, Etc., Government Printer, Wellington, p. 489.

New Zealand Official Year Book 1924, Section XV, Land Tenure, Government Printer, Wellington, p. 356.

New Zealand Official Year Book 1941, Section XVII, Land Tenure, Government Printer, Wellington, p. 317.

Roche, M 2008, "Failure deconstructed: histories and geographies of soldier settlement in New Zealand circa 1917–39", *New Zealand Geographer*, Vol. 64, No. 1, pp. 46–56.

Rutherfurd, J to Commissioner of Crown Lands, 1920, Section 16, Block XI, Retaruke, SD AAMA Acc W4320 36 RL203 WJHShaw Sec 3, Blk XIV, Retaruke, Archives New Zealand, Wellington.

Rutherfurd, J to Commissioner of Crown Lands, 14 March 1922, AAMA Acc W4320 36 RL203 WJHShaw Sec 3, Blk XIV, Retaruke, Archives New Zealand, Wellington.

Rutherfurd, J to Commissioner of Crown Lands, WJH Shaw, 6 January 1927, AAMA Acc W4320 36 RL203 WJHShaw Sec 3, Blk XIV, Retaruke, Archives New Zealand, Wellington.

Shaw, William James Henry Mitchell – WW121429 – Army, AABK (New Zealand Defence Force, Personnel Archives) 18805 W5553 13/0103785, Archives New Zealand, Wellington.

Shaw, WJ to Commissioner of Crown Lands, 30 April 1923, AAMA Acc W4320 36 RL203 WJHShaw Sec 3, Blk XIV, Retaruke, Archives New Zealand, Wellington.

Shaw, WJ to Commissioner of Crown Lands, 24 April 1929, AAMA Acc W4320 36 RL203 WJHShaw Sec 3, Blk XIV, Retaruke, Archives New Zealand, Wellington.

Statistics of New Zealand 1911, Occupied Lands, Government Printer, Wellington, p. 514.

Williams, DO 1936, "Land settlement and finance", in H Belshaw, DO Williams, FB Stephens, EJ Fawcett and HR Rodwell, (Eds) *Agricultural Organisation in New Zealand*, New Zealand Institute of Pacific Relations & Melbourne University Press, Melbourne, pp. 123–49.

Winder, G 2009, "Grasslands revolutions in New Zealand: disaggregating a national story", *New Zealand Geographer*, Vol. 65, No. 3, pp. 187–200.

Wright Stephenson and Co. to Commissioner of Crown Lands, 12 July 1922, AAMA Acc W4320 36 RL203 WJHShaw Sec 3, Blk XIV, Retaruke, Archives New Zealand, Wellington.

6 Western Australia's interwar group settlement scheme, a case study of the Leeds Group, 1923–1938

Mark Brayshay

Introduction

Weakened by the demands of the Great War and the dislocation of its overseas trade markets, Britain's economy went into a recession which, by the end of 1920, had triggered a sharp rise in unemployment. Within a year, official statistics showed that 17% of the insured labour force[1] was out of work (Constantine 1980, pp. 1–16; 1984, pp. 89–93). An economic analyst at the time pointed to the disproportionately high representation of males in their twenties among the jobless (Morley 1922, pp. 487–8). The distress experienced by a group that included many ex-servicemen and their families inevitably evoked a sense of guilt that those whose efforts on behalf of the nation had been vital in the conflict faced so bleak a homecoming (Powell 1981, pp. 64–87; Fedorowich 1990, pp. 51–4; Perry 2000, pp. 58–80).[2] Although some who had served in the forces received small pensions, for most of the unwaged, life depended on dole money or on poor-relief payments. Against the background of these circumstances, and amidst calls at both national and local levels for skills training for returning soldiers, there was a renewed focus on assisted emigration to Britain's empire.[3]

Annual emigration from the United Kingdom had averaged 260,000 in the years before 1914 but, during the war, the outflow dwindled virtually to zero (Pope 1977, pp. 194–5; Langfield 1999, pp. 55–6). However, emigration possibilities for ex-servicemen were being actively explored by the British Government's Reconstruction Committee within a few months of its creation in 1916. A year later, an Empire Settlement Committee was established to formulate new policies for resettling demobilised soldiers and sailors in Britain's dominions. Reconstituted as the Oversea Settlement Committee (OSC) in January 1919, and chaired by self-confessed empire-enthusiast Leo Amery, its work led to a scheme for the provision to eligible ex-servicemen of free-passages to the four self-governing dominions: Australia, Canada, New Zealand and South Africa (Amery 1953, Vol. ii). For the first time since the early nineteenth century, Britain was thus committed to state-aided imperial emigration (Kennedy 1988, p. 406). Meshed with soldier settlement and employment programmes devised by each dominion government (see Chapter 5 on New Zealand by Michael Roche), the OSC's scheme ultimately involved around 86,000 participants (Powell 1981, pp. 66–80; Kennedy 1988, p. 415; Johnson 2005, pp. 496–9). The total was well below the 450,000 migrants initially envisioned (Fedorowich 1990, pp. 49–65).[4] Under-resourced and poorly organised, soldier settlement was not an unqualified success, but the experience gained informed debates about wider programmes of assisted emigration as a way to alleviate home unemployment (Selwood and Brayshay 1998, pp. 480–2).

In September 1921, Amery urged the British Cabinet to support a new law to commit government financial assistance for *all* needy emigrants.[5] As the costs of unemployment relief rose, the argument that subsidies for emigration would not only be less expensive in the longer term but also offer hope to the jobless and economic benefits to the Empire proved highly persuasive (Williams 1990, pp. 22–40). In due course, the Empire Settlement Act of 31 May 1922 was passed and committed Britain to spend £3 million annually for 14 years on approved emigration and settlement schemes in the dominions (Drummond 1974, p. 43).[6]

As the Empire Settlement Bill progressed through Parliament in Westminster, Western Australia's flamboyant premier Sir James ("Moo Cow", as he was termed in the local press) Mitchell toured England and Scotland to promote the "strong bonds of Empire", the notion of a "white Australia" and a significant increase in migration to his State.[7] Mitchell told London audiences that Western Australia comprised 624 million acres but had a population of only 330,000. There was room, he said, for "millions more", to exploit the State's rich natural resources.[8] Between visits to Birmingham, Nottingham, Bradford, Manchester and Glasgow, Mitchell held talks with the OSC and consulted William Hughes, Australia's Commonwealth Prime Minister, by telegraph and letter. With serendipitous timing, in late June 1922, just before his departure from England, Mitchell theatrically announced that Western Australia would be the first to take advantage of the newly minted "Imperial Act" because a "tentative agreement" had been reached whereby his government, the Commonwealth (Federal) Government of Australia, and the British Government would jointly support a major expansion of the "group settlement scheme" being piloted in his State.[9]

When eventually agreed upon, the "business model" for the establishment of profitable agriculture in the State's South West envisaged the provision for each new settler of a freehold farm of up to 160 acres on Crown Lands and a loan of £1,000 to pay for assisted passages, agricultural development costs and the subsistence wages payable while preparatory work occurred. The capital required was to be raised by borrowing and interest payments shared equally for five years by the three participating governments. Thereafter, it was expected that settlers' farms would be sufficiently productive to enable them to begin repaying their loan and interest debts, though ultimate liabilities were, in any case, to be transferred solely to Western Australia and its government-owned Agricultural Bank.

Group settlement in Western Australia

From its inception, Western Australia's group settlement scheme has attracted much scholarly attention (Morrison 1923; Shann 1925; Hunt 1958; Gabbedy 1988a, 1988b; Bolton 1994; Roe 1990, 1995; Brayshay and Selwood 1997a; Crawford and Crawford 2003; Jones et al. 2015). Moreover, it has not only bequeathed a recognisable landscape legacy but has also become embedded in the State's overall social and political history. However, in the context of Australia as a whole, interwar rural settlement was unusual: a majority of immigrants in that period – overwhelmingly British – were not "rural pioneers" but tended instead to enter the urban working classes (Jupp 1990, pp. 285–91). Countering that norm, the scale and soaring ambition of Western Australia's group settlement, as well as its manifold failings and later notoriety, set it apart as a paradigm of twentieth-century organised rural settlement.

Before the Great War, Western Australia experienced a short-lived, modest surge in demographic and economic growth. Fuelled by a gold rush beginning in the early 1890s, it was largely spent by 1910. An urgent need for renewed development impetus, based on more diversified wealth creation, therefore existed. Attention turned to the potential for agricultural expansion in the vast, sparsely populated, forested tracts of land in the State's South West (Jones et al. 2015, p. 126). Areas of land where blocks (farm holdings) might in the future be established were surveyed and mapped. It was recognised that the inevitably heavy work involved in clearing forested land for agriculture would demand well-organised efforts by large "armies" of would-be farmers.

While some elements of the idea of undertaking such work by means of collaborating groups of pioneers had been tried already in the programme for resettling ex-servicemen, Premier Mitchell's first announcement of his intention to inaugurate a general group settlement scheme came a few months before his 1922 tour of Great Britain.[10] His pilot plan called for potential farmers to apply to the State's Agricultural Bank to take part in a trial development of Crown Land near Manjimup, where 20 blocks had been surveyed and the soils declared to be of good quality, "suited for dairying, potato and fruit growing". Teams of men would clear core areas of each block, erect fencing, create water supplies and construct dwellings. Once the preparation was complete, the blocks would be ready to allocate by ballot to participants (Blond 1987, p. 2; Brayshay and Selwood 1997a, pp. 201–2).[11]

Although later State premiers, as well as senior ministers, played roles in administering (and reforming) group settlement, Mitchell has been credited as the originator of the idea; he did little to disabuse the attribution. However, it is now clear that all the scheme's ingredients were set out in 1917 by Ernest G. Marlow in a proposal entitled: "A scheme to settle returned soldiers and others on the land" (Selwood and Brayshay 1998, pp. 408–10).[12] The Western Australian government "cherry-picked" Marlow's ideas for soldier settlement, and Mitchell comprehensively quarried them in 1921 for his group settlement project. Although the earliest settlers came from within Western Australia, the numbers available were insufficient, and Mitchell therefore turned his attention to immigration from Britain. Between the first trial of the project in 1921 and late 1928, some 6,000 settlers, predominantly new arrivals from Britain, were placed across 127 locations termed "Groups", where they cleared 98,800 acres of forest in the State's South West (Figure 6.1).[13]

It is clear that the full-scale launch of group settlement, occurring after Mitchell's 1922 return to Perth and the signing of the tripartite government agreement on sharing the scheme's start-up financing, depended critically on (1) the confluence of unusual political, economic and demographic circumstances prevailing in the United Kingdom and Western Australia after the First World War; (2) experience gained in attempting to place groups of ex-servicemen on undeveloped rural land; (3) post-war gratitude in Britain for the dominions' support during the conflict that, in some quarters, rekindled an interest in the Empire and (4) the personal commitment and zeal of Sir James Mitchell. The coincidence of these four factors gave the scheme considerable momentum irrespective of whether its provisions were realistic or workable. By the late 1920s, however, few of those involved in the State's group settlement project could escape its shortcomings and failures.

The Leeds group settlers

It is largely upon the testimony of the triumphs and difficulties experienced by one particular contingent of British group settlers, migrating together from the city of Leeds and its immediate environs that this chapter principally draws. The story of the original Leeds Group, comprising only 19 families (79 individuals), encapsulates the story of the whole group settlement episode in early twentieth-century Western Australia and also evinces their contribution to the making of a distinctive landscape in remote rural forest and bushland.

While typical of group settlers across the State's South West, the Leeds company differed starkly from the majority in one respect. Instead of their party being assembled, almost at random, as was usually the case, from among British migrants who arrived in Western Australia and lodged temporarily in the Immigrants Home in Fremantle, the Leeds band was already formed as an entity at home before departure. Only the small parties pre-formed in Devon and Cornwall, Catterick in Yorkshire and Greenock in Scotland were similarly drawn from specific geographical parts of Britain (Brayshay and Selwood 1997a, 1997b; Selwood and Brayshay 1998).[14]

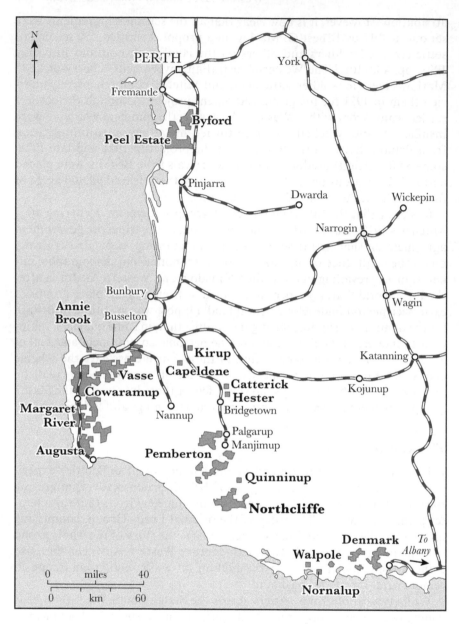

Figure 6.1 Western Australia's Group Settlement Areas by the late 1920s.

Based on: Gabbedy, JP 1988, *Group Settlement, Part 1: Its Origins: Politics and Administration*, University of Western Australia Press, Nedlands, p. ii

Although their common origin ultimately did nothing to help Leeds settlers to clear forests and cultivate (sometimes hopelessly infertile) soils as
pasture for dairy cattle, or face the burden of paying down debts or cope
with the collapse in agricultural commodity prices wrought by the Great
Depression, their own testimony indicates that their shared background
forged a lasting social bond; as one emigrant later recalled, "One thing
holding things together was that they all came from Leeds; it lent a 'spirit
to the group'".[15]

The cohesion of the Leeds migrants owed much to the initial formation
of the party in 1923 by Major (later Sir[16]) John Dearman Birchall, one of
the city's members of the UK parliament. For the next 15 years, Birchall
provided support, followed progress and maintained contacts with the settlers. Enlisting help from several Leeds organisations, he created networks
of links that reinforced attachments to their old home community. Birchall
also used his influence to champion their cause in Britain and with Western
Australia's authorities. Over the same period, the local Yorkshire press sustained an ongoing interest in the migrants; and, in due course, the city's
Anglican Church and other local benefactors, at Birchall's request, finan-
cially supported Christian ministry in Northcliffe: the parish to which the
Leeds people went. The MP's ongoing commitment to his "Leeds pioneers"
is certainly a striking aspect of their story; Birchall's support boosted their
morale during challenging times and kept alive lasting links with the city
from which they came.[17]

Although evidence of the Leeds migrants from both the private archives
of Sir John Birchall and the Yorkshire newspapers has previously been cited
(Gabbedy 1988b; Crawford and Crawford 2003; Perry 2012, 2014), in this
chapter, through the lens of the Leeds case study, the material is employed
specifically to illustrate and understand the human experience of Western
Australia's rural settlement experiment.

A "New Leeds" in Australia

During 1922, when unemployment rates in Leeds grew to be among the
highest in the country, Birchall formulated his ideas for a "New Leeds"
overseas.[18] He obtained a copy of the new Empire Settlement Act, examined
literature describing opportunities in Canada and sought press cuttings of
reports of Sir James Mitchell's speeches in Britain. He was impressed by
arguments, advanced by Malcolm Bruce during the late-1922 Australian
federal elections, advocating migration as a means of tackling Britain's
overpopulation and unemployment and Australia's shortage of manpower.[19]
He obtained from Australia House in the Strand copies of (1) the Western
Australia Group Settlement Conditions, (2) the member's agreement and
(3) the draft permit to occupy holdings. He was also sent a copy of the illustrated brochure describing progress on the reclamation, development and
group settlement of the Peel Estate, an area lying to the south of Perth,

which Mitchell used to showcase the scheme (*Group Settlement in Western Australia* 1923). Weighing the relative merits of various emigration opportunities spelt out by the OSC (1923), Birchall was most persuaded by the Western Australian proposals for British families with little or no start-up capital to become pioneer battalions, carving out new farms on Crown Land. Ideas for Devon and Cornwall group migration, published at the time by Lieutenant Colonel Stewart Newcombe, chimed closely with Birchall's thinking and helped him to crystallise his plan (Devon and Cornwall Migration Committee 1923).[20]

In October 1923, Birchall announced his proposal in full for a "Leeds colony in Australia". The *Yorkshire Evening Post* publicised the MP's call for "families, or single men or women, willing to go from the city" to send him their names, addresses, occupations and numbers in families.[21] The idea immediately attracted widespread attention. Britain's national newspapers carried the story, and it soon reached Australia where, for example, on 17 October, it appeared in the (Brisbane) *Courier*, which added Birchall's comment,

> I find that the fear of loneliness and isolation is the greatest objection to the old methods of immigration. Now I am told officially...that arrangements will be made for whole families to migrate together...for groups from the same place in the Old Country to be kept together in the new lands.[22]

While there was plentiful support for Birchall at home, an anonymous objector wrote him a nasty letter, and rather more vented their negative reactions in letters to newspapers. Supporters of the scheme were accused of sending out "sound Yorkshiremen" to Australia to work for "next to nothing" to enrich another part of the world, while at home they were being replaced by too many foreigners: the "scum of Europe", entering via England's east coast. Birchall was told to "wait outside Armley Gaol" and instead recruit released prisoners to go to Australia.[23] However, within days, the MP reported the receipt of around 500 positive "expressions of interest".[24] On 26 October, interviews were held at the Leeds Employment Exchange, supervised by JT Barnes (Australian Deputy-Director of Migration), who travelled from London for the event. While only 122 applicants attended the interviews, all were instructed to bring their wives and children, and more than 400 people therefore crowded into the building. "Every other young married woman had a child upon her knee; there were scores of tiny toddlers and several babies under a year old".[25] Birchall later annotated press cuttings reporting the event: "1st interview, 122 came, 60 rejected, 24 reserved, 31 provisionally accepted for group, 7 for other schemes (mainly Victoria)".[26]

After the interviews, Birchall hosted a dinner at the Great Northern Hotel for a dozen dignitaries and officials including JT Barnes; the Lord Mayor,

Frank Fountain; Shipping and Emigration Agent, CA Hood; and Migration Officer, North-Eastern Division of Leeds, Captain Frank S. Baker.[27] There was confidence among the diners that they were launching a project that could potentially do a great deal of good both for Leeds and the Empire.

A hiatus thereafter occurred as assurances were sought that the Leeds party would be placed *together* in Western Australia and not dispersed across several group settlement locations. In his letter of 4 December to prospective emigrants, Birchall confirmed receipt of an affirmative response from Perth that migrants recruited in Leeds would be "settled in the same locality". Confirmation also came from London's Australia House that 20 Leeds families, subject to approval, would definitely be "kept together" on the "same terms as those agreed for Devon and Cornwall". The would-be Leeds settlers were asked to make their final firm commitment.[28]

Emigration

Birchall enlisted the Yorkshire Ladies' Council of Education to arrange personal "liaison" visits to each emigrant family to ascertain their readiness and whether they needed specific extra items, such as warm clothing or shoes. He encouraged the forming of correspondence links between the Ladies' Council and female emigrants.[29] He also contacted the Church Army's Overseas Settlement, Passenger and Shipping Bureau to select suitable books for Leeds emigrants to read during their journey; some packages were later made up, at Birchall's expense, by Edward North's bookshop in Paternoster Row.[30]

By February 1924, the final list of 19 (not 20) families was ready. Just six of the men possessed experience at farm work, but two more claimed to have "knowledge of working with horses", and another was a gardener. The employment skills of the remaining ten included engineering, joinery, metalworking, motor vehicle driving, electricity cable-laying, bar work and the police force. A majority had fought in the Great War, and some were armed forces veterans with a service record of ten or more years. The men ranged in age from those in their early 20s, such as Frederick Bean, to a few who had reached their 40s, including Joseph Brearley and Albert Bristow. Before emigrating, the last Yorkshire addresses of the party indicate that 15 of the families had resided in the city itself, a suburb, or a very close neighbouring village. One family was from the more distant small town of Dewsbury, and another three came from relatively rural communities, namely Guiseley, Castleford and Fulford (Table 6.1). Unfortunately, evidence of the previous work experience of wives and older children is very sketchy. Several women appear to have been engaged in "domestic" work, but Herbert Marriott's wife stated, in a later letter, that she had formerly been employed in a city print works.[31]

The Lord Mayor hosted a civic farewell at which each of the Leeds Group of migrants received a small gift from the city, and the women were

Table 6.1 The 19 original Leeds settler families going to Northcliffe Group 107, Western Australia, 1924

Names of settlers (& wife, if known)	Settler's occupation or skill, before emigration	Last known Leeds/Yorkshire address	Block	Children (by 1926)	Year left scheme
Bayliss	Herbert S.	Farm & horse work	24 Houghley Grove, Armley, Leeds	9013	one son	1925
Bean	Frederick & Edna	Motor car driver (knowledge of horses)	43 Clifton Grove, Harehills, Leeds	9018	one daughter	1937
Brearley	Joseph & Alice	Joiner (knowledge of horses while in Army)	21 Cross Green Crescent, Leeds	9008	two sons and one daughter	1930
Bristow	Albert Henry & Queenie	Electric light company van driver	27 Fir Parade, Ravensthorpe, Dewsbury	9010	two sons and two daughters	1932
Downs	John T. N.	Farm work	Main Street, Shadwell, Leeds	9007	one son	1928*
Foxcroft	Edgar	Royal Navy 14-years; barman (father's pub)	54 Bayswater Avenue, Roundhay Road, Leeds	9135	one son and one daughter	1927
Gallimore	Ernest William	Farm work	Norland Farm, Carlton Lane, Guiseley	9012	one son and one daughter	1925?
Grimshaw	Walter & Elizabeth	Police constable	16 Sugarwell Mount, Meanwood, Leeds	9006	one son and one daughter	1940s?
Jacobs	C. James & Mary Jane	Collier; formerly farm work	14 Ledsham Terrace, Hunslet, Leeds	9019	two daughters	1928*
Johnson	George William & Annie	Steel worker	7 Minnie Street, Burley, Leeds	9004	one son and one daughter	1926
Leese	John T.	Farm & horse work	10 Henson Grove, Airedale, Castleford	9133	two sons	1933**
Marriott	Herbert S.	Engineer	61 Whingate Avenue, Armley, Leeds	9000	two daughters and one son	1925

Moody	George F. & Elizabeth Lucy	Electricity cable layer, recently returned from India	12 Acre Square, Middleton, near Leeds	9003	one daughter	1930s?
Palmer	John & Laura	Farm work	Swiss Cottage, Fulford, near York	9134	two sons and five daughters	1940s?
Parish	Albert Henry	Foundry hammer man	6 Back Burley Street, Leeds	9011	one son	1928
Smith	John William	Labourer	16 Bridgewater Terrace, Hunslet, Leeds	9010	?	1924
Smith	Wilfred	Stoker (formerly in Royal Navy)	4 Sheffield Street, Ellerby Road, Leeds	9005	two sons and one daughter	No info.
Wilson	George	Excavator (building trade?)	13 Saville Yard, West Street, Leeds	9015	one son and one daughter	1927
Wilson	Herbert E.	Gardener	11 Huttons Row, Green Road, Leeds	9014	two sons	No info.

Notes: A majority of the men were demobilised soldiers or sailors who served in the Great War; some had been in the Armed Forces for longer. While 'occupations or skills' before emigration are known, whether or not they were in secure work is not clear. Block numbers refer to the holdings first allocated by ballot to the settlers in Group 107. The year shown in the final column indicates when a family left Group 107. The numbers of children are taken from figures published in the reports of J B Lillis, 1926.

*C. James Jacobs was moved to Group 148, but later left the scheme.
**John Leese was moved twice to other groups, but left the scheme entirely in 1933.

presented with another by Mrs. Birchall.[32] Wearing his military medals, former city police constable, Walter Grimshaw, thanked Leeds on behalf of his fellow emigrants.[33] Birchall received confirmation that passage aboard the SS *Diogenes*, departing from Tilbury on 1 March, was arranged and that overnight rail travel from Leeds to King's Cross could be booked; the group would then catch another train from St Pancras to the docks. On their arrival in London, Major and Mrs Birchall met the emigrants and paid for the group's breakfast prior to accompanying them to the embarkation.[34]

Arriving on Group 107, Northcliffe, Western Australia

Walter Grimshaw fulfilled his promise made before departure to send accounts to the *Yorkshire Post* of the Leeds Group's journey, via Cape Town, to Albany, arriving on 6 April 1924.[35] Travelling thus far with the Leeds party, Devon and Cornwall's (second) group bade farewell to them and was transferred to its destination in the nearby town of Denmark (Brayshay and Selwood 1997a, pp. 214–5). The Leeds settlers boarded a train for Fremantle and spent a week lodging in the "Immigrants Depot", which they said was "alive with bugs" (Daubney 2001, p. 11). Taking another train journey, the settlers journeyed back to Pemberton, sleeping overnight in the carriages before moving onwards in trucks along bush tracks to their remote campsite on Group 107, 12 miles west of Northcliffe (Figure 6.2).[36] Only ten galvanised, corrugated-iron, two-roomed shacks had been erected and a shallow well sunk, which yielded scarcely potable water. The shacks had no floors or windows, only doorways covered with sacking. Eighteen of the families were initially required to share a hut.[37] The exceptionally Spartan conditions were a shock, especially to the womenfolk. Immediate tasks included erecting another nine shacks and deepening the well.

Birchall had been promised that provision would be made for the education of the children, healthcare and Christian ministry, but these amenities were not immediately available. The dirt track into Northcliffe meant an arduous journey to obtain supplies, and freight charges added considerably to the price of goods. The settlers were assured that, within nine or ten months of work commencing on clearing the land, each family would move into a government-supplied, timber-built, four-roomed cottage on their individual property. Although virtually all members of the Leeds party had indeed, by early 1925, transferred from the campsite to their own locations, no cottages were yet built. The metal shacks ("humpys") had instead been dismantled and carted onto the various farms for re-erection.[38] Some had attempted to make the huts more "comfortable". Obtaining supplies from Perth, Joe Brearley fitted glass in windows he had cut into the walls and fixed asbestos sheeting as cladding to the exterior, declaring the latter "a decided improvement on corrugated iron".[39] It was April 1926, two years after the Leeds Group arrived, before timber for "our bungalows" was being

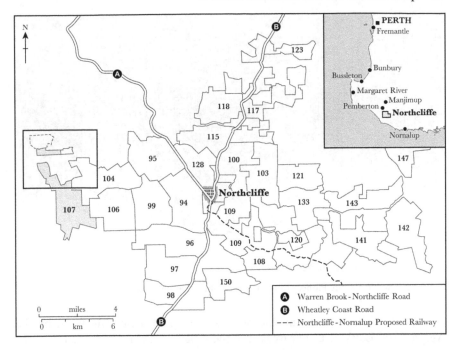

Figure 6.2 The Northcliffe Groups, circa 1928. Group 107 was settled by the emigrants from Leeds (shaded) and the framed section is that depicted in Figure 6.3.

Adapted from: Gabbedy, JP 1988, *Group Settlement, Part 2: Its People: Their Life and Times – An Inside View*, University of Western Australia Press, Nedlands, p. xviii

freighted out in motor trucks. The settlers were promised that "we shall have them up before this winter", but Brearley thought that the contractor would "have to get a move on, or the roads will be too bad to get the timber out [here]".[40]

Clearing the land

Group 107 comprised a heavily wooded area of 2,500 acres. Its most westerly point was only five miles from the Southern Ocean. Removing huge eucalypts, including karri, marri (red gum) and jarrah, and clearing grass-gum trees standing amidst dense scrub, proved formidable tasks. The settlers' work was supervised by their officially appointed foreman, Arthur "Dolly" Twigg. In common with others holding similar posts elsewhere, Twigg's expertise was rather limited; moreover, at various times, he proved inept in managing the settlers and dissatisfaction occurred.[41] His crucial task was to select the location of the 25-acre core areas to be cleared for each holding. Twigg appears frequently to have picked spots that were more easily tackled rather than those with the best soils or topography.

Figure 6.3 The Group 107 campsite, 1924.

Courtesy: The Northcliffe Pioneer Museum, The Palmer Family Archives

While working as a team, the settlers were paid a flat-rate wage by the Agricultural Bank of £3 per week, but the scheme was later changed, and the men were placed on individual contracts and earned income in accordance with the amount of work that they were deemed (by their foreman) to have done. However, whether a settler was paid a fixed rate weekly wage or was placed on a contract, all sums paid out as subsistence wages were added to his accumulating debt, owed to the Agricultural Bank for the development of his farm, which he was legally liable to repay once the holding became a commercial dairy farming enterprise. A cart and three horses were provided for the camp to share for transport and work on the land (Figure 6.3). Because there was, initially, no natural grazing, horse fodder costs were also added as "administrative charges" to the settlers' growing debts (Gabbedy 1988b, p. 243). Smaller trees, with circumferences up to 6.5 feet, were uprooted using cables and horsepower. Sawn into pieces, the timber was burnt. Large trees were tackled either by ring-barking (sap-ringing) and being left to die or by destroying their roots with explosives. The latter approach was expensive and hazardous but was thought, at the time, to suppress the regrowth of suckers (*Report of the Royal Commission on Group Settlement* 1925, 28, p. 811).[42] Leeds settlers reported tackling trees with trunks exceeding 30 feet in circumference. Gelignite shattered main roots but often failed to bring trees down; it was necessary to set them ablaze and hope the wind would topple the remnants.[43] Neither sap-ringing nor gelignite, in fact, prevented the growth of innumerable suckers, which required repeated grubbing out, usually by hand.

Mattocks, spades and axes were used to remove lighter scrub, but unrelenting vegetation regrowth nevertheless demanded repeated burning off[44]; and some Leeds settlers struggled for years to establish a sustainable sward on their sandy, acidic soils. Waterlogged in winter, such poor soils were characteristic of tracts of the "plain country", which was common on Group 107.[45] It was later alleged that the need in many of the groups for comprehensive land drainage, and the unsuitability of the soils for dairy farming, were well known in advance by promoters of the scheme.[46] Pasture simply did not thrive without repeated top dressing that, in the absence of sufficient cattle to produce natural fertiliser, added significantly to costs. Several Leeds settlers said they repeatedly needed to recondition their pastures by uprooting tree suckers, ploughing and re-seeding.[47]

During their first year in Australia, despite the physical demands of the work for both men and women, the mood of the Leeds Group was generally optimistic. Heartened to receive Christmas greetings and a calendar from Birchall, John Leese announced the arrival of his second son: "Life ... is jolly good ... it's a bit rough, but not half as rough as I expected". Leese felt that there were "thousands of acres of good land waiting for somebody to come along and adopt it" and going to the group settlement had been "the finest step I ever took in my life, and I am now just as determined to stick it and pull through".[48]

A school, medical services, progress on
the farms and early departures

Some 16 months after the Leeds settlers arrived, a "double school" was built to serve Groups 104, 106 and 107. Opening in August 1925, Frank Smith and his wife became teacher and assistant teacher.[49] For most Leeds children, attendance meant a walk or a ride by horse and cart of three miles. Though not all had yet reached school age, there were 39 children on Group 107 (Lillis 1926a, p. 6). Healthcare provision in Northcliffe by Dr Lionel Ward greatly benefited Group 107.[50] His main task was, he said, "ushering little Australians into the world" (Perry 2012; Lillis 1926b, p. 8).

Walter Grimshaw told Birchall that the group's men were busy "burning off" land to make it ready for sowing, and he sent thanks for the medicine chest supplied for communal use by the MP.[51] The mood was still positive in October 1925 when Mary Jacobs sought Birchall's help to bring out her son-in-law, daughter and two grandchildren to Group 107.[52] However, by then, three of the original 19 settlers had left. In two cases, personal reasons had caused them to depart.[53] The third to leave, Herbert Marriott, although injured when tree felling on 16 April 1924, had recovered well enough to return to work within a few weeks, and the family was apparently still optimistic about the prospects for their farm when his wife wrote to Mrs Birchall on 29 January 1925.[54] However, by November, the family had decided to abandon the group settlement scheme entirely and had moved to Perth. Marriott later explained that the debt that he would ultimately owe for the costs of developing his holding was likely to far exceed the original estimate of £1,000, and "I saw that if I continued there much longer, I should be up to my neck in debt for the rest of my life".[55] Obtaining employment in Perth in his former trade as a mechanic, he paid back what he owed to the Agricultural Bank and then saved £167 to fund the family's return to England. Their homeward journey aboard the SS *Themistocles* brought them to Southampton on 22 June 1927. At the time, Birchall was attempting to form a second Leeds Group to go to Western Australia, and he was extremely anxious to dampen the bad publicity that had been triggered both in the city and elsewhere by Marriott's headline-grabbing homecoming.[56]

Though critical of some of his fellow settlers who he said were workshy, Walter Grimshaw acknowledged that "in the first year, things were terribly hard for everybody, the work was new and all of us suffered much with raw hands and raw tempers, which life in the shacks during wet weather did nothing to mitigate".[57] While other settlers, placed on the worst of the sandy land, were doing noticeably less well, both Grimshaw and Joseph Brearley claimed to be obtaining crops of maize, potatoes, other vegetables and fruit, as well as sowing 25 acres of clovers and grasses (Figure 6.4). Initially, at least one cow was allocated per settler to provide for his own family's needs and to allow him to gain experience in handling livestock; they were also supplied with their own horse and cart (Gabbedy 1988b, p. 243; Crawford and Crawford 2003, pp. 94–5).[58] Although more pasture and hay crops were gradually produced, very few holdings ever yielded enough, and the cost of brought-in feed was an unanticipated burden. Group foremen

Figure 6.4 Walter and Elizabeth Grimshaw, scything hay, Group 107, circa 1929.

Courtesy: The Northcliffe Pioneer Museum, NH35-16

were empowered to repossess for "reconditioning" any cattle deemed to be underfed and weak.[59] Gradually, however, some Leeds settlers on Group 107 increased their herd and eventually produced cream for commercial butter production in the factory at Bunbury.

Before then, links with Leeds were strengthened when, in January 1926, *Yorkshire Evening Post* reporter J. B. Lillis visited Group 107 to gauge how the settlers were progressing. Lillis interviewed almost all of the 16 remaining families, and his articles describing their lives were later published (Lillis 1926a–c). Noting the frustration of many regarding the lack of the promised bungalows and the sheer scale of the task of land clearance, the mood of confidence still persisted. In addition to their horse and single cow, families were keeping pigs and poultry, and a few had created vegetable gardens. Road gangs, boarding with the settlers and paid wages by the State, had begun the much-needed improvement of the road to Northcliffe.[60] Building the 30-mile road that connected Northcliffe and Pemberton was also in progress.

In October and November 1925, William Bankes Amery, the British government representative for migration in Australia, toured the group settlements. He took particular interest in the Leeds Group but actually spent only a few hours in their settlement, allowing time to interview just three families: the Bristows, Foxcrofts and (Albert) Wilsons. Concluding that the well-being of the womenfolk was critical to the success of the scheme, Bankes Amery recorded their feelings of isolation in such a remote location. Many shared the heavy farm work with the men, partly to combat their loneliness. Dances and other social events held in the schoolroom were therefore very popular as opportunities for community interaction. However, he noted that there had been fewer quarrels amongst the Leeds settlers than had occurred elsewhere.

Offered no definitive information on the matter, settlers expressed concern regarding their probable final indebtedness and were worried about the likely rate of interest that they might face (Bankes Amery 1926). The same disquiet featured prominently in a minority report by Charles Latham appended to the 1925 *Report of the Royal Commission on Group Settlement* (pp. xvii–xviii).[61] The government's response was to pass enabling legislation on 31 December 1925 to empower the State's Agricultural Bank to take over the full management of the group settlement scheme. It was also recognised in the report that, in order to "reconstruct" or merge holdings to achieve farms of a size capable of being profitable, the originally specified upper limit of 160 acres would have to be breached.

Reconstruction

The reconstruction process began in 1927. Overseen by the Minister for Agriculture, Frank Troy, it required (1) identification of blocks (sometimes entire areas) to be condemned as completely unsuitable for cultivation and

(2) amalgamation of those blocks deemed to have farming potential so as to form larger holdings. There was, of course, often a heavy human cost. Reconstruction did not affect Group 107 until 1928, but the mere threat of it contributed a year earlier to the decisions of Edgar Foxcroft, George Johnson and George Wilson to retire from the scheme.[62]

For those directly impacted, the experience of relocation could be traumatic. Some who were sent to distant groups were, in the short term, distressed and destabilised. One example was settler John Leese. To form a single larger farm by merging the properties of Leese and his neighbour, Jack Palmer, one of them had to relocate. In making the choice, much apparently depended on the relative size of the two affected families: Leese, his wife and two boys were moved (to Group 117 and later to Group 2), while Palmer's household, comprising ten children, remained on Group 107 (Figure 6.5).

Mildred Ormerod of the Yorkshire Ladies' Council for Education received an anguished letter from Mrs Leese, complaining about her family's ejection from their first farm:

> ...We have slaved day and night to make the place into a paying concern; spent many pounds of our hard-earned money in pulling the place into what it is today, to start all over again on new ground, just because we was not pals with the Foreman and buy him whisky, etc., which was his favourite meal every day ... On the whole, we are very discontented on our new block, as there is no place like home and the Leeds Group is our proper home, for all our old friends are there.[63]

Ormerod sought Birchall's advice and he brought the case to the attention of William Angwin, Agent General for Western Australia in London. Angwin in turn referred the matter to his government in Perth.[64] Birchall's archives offer a glimpse of the bullying of some settlers by Western Australia's group settlement authorities. The decision to move the Leese family had, it appears, been made by the field supervisor not the foreman; though the latter was reportedly "replaced".[65] Contrary to Mrs Leese's claim, the field supervisor reported that the original holding was not a "paying concern" but was instead "very poor". John Leese was questioned and had, apparently, not supported his wife's statements. On the contrary, allegedly, he was content to move. In winding up the affair, Mrs Ormerod perceptively remarked that Leese was perhaps "anxious to keep on good terms with the authorities on the spot" and she hoped that Mrs Leese's "rather injudiciously worded letter" would "not do him any harm".[66]

George Moody's holding on Group 107 was enlarged in the reconstruction exercise, but he appears to have struggled to cope. He complained to Birchall in June 1929 about the new foreman, Jim Roche, who had replaced the dismissed Arthur Twigg. On Roche's orders, Moody was denied his contract pay that month because, it was said, he had failed to recondition his

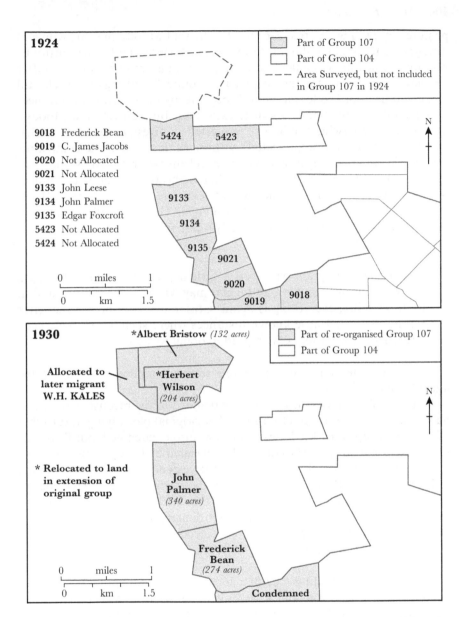

Figure 6.5 The 1928 reconstruction of nine blocks in part of the Group 107 settlement (see Fig. 6.2) to form five larger farms. As well as the blocks abandoned by Foxcroft and vacated by the relocation of Leese, an area of previously surveyed but unused blocks and four unallocated blocks was incorporated in the reconfiguration of this small area of the settlement.

Courtesy: The Northcliffe Pioneer Museum, The Palmer Family Archives pastures by pulling out tree suckers. Roche told Moody that his wife and nine-year-old daughter should do the work. Angered by Roche's attitude, Moody claimed he was unable to manage on an average net income (after deductions for the doctor and hospital) of around 47 shillings per week, still less could he afford to lose a month's pay. He also documented grievances of some of his neighbours. As in Leese's case, Birchall referred Moody's complaints to Angwin, who in turn requested an investigation in Australia. Again, the matter was adjudicated by those against whom the complaints were made, and the allegations were comprehensively rejected. Moody was questioned and forced to name the neighbours he alleged had been bullied, but none of the parties concerned would acknowledge any grievance. Even Birchall was obliquely admonished by the Agent General, William Angwin. Forwarding written responses to Moody's case from Western Australia, he patronisingly remarked that,

> ...A settler or migrant who feels a little depressed owing to having a difference of opinion with an official may write ... at the time, and in all probability, everything is again satisfactory long before the letter reaches here ... I have known instances where such letters have been written and the writers have regretted doing so later.[67]

Birchall thereafter forwarded no more settler complaints to any of the agents general.

James Jacobs's block was another condemned in the reconstruction, and he was sent to a holding on Group 148. The original block lying next to that of Jacobs, occupied by Frederick Bean, was also condemned, but Bean was relocated *within* Group 107 to the holding vacated by Foxcroft, plus another two, as yet, uncleared holdings. His farm then extended to 274 acres. Writing to Birchall in 1931, Bean said, "I am one of the original settlers that came out with this group from Leeds although I am ashamed to say I am probably the only one who has not written to you, regarding my progress". Remarkably, Bean had overcome the upheaval of reconstruction and the task of clearing yet more land. He reported that,

> ...Our present holding (our first location was condemned) comprises mostly first-class land and 40 acres is under pasture, which is excellent. I cut 30 tons of clover hay from 7½ acres last season. I have other land, which is partly cleared and I will put it under cultivation or pasture next year. Our stock consists of 11 milking cows (9 of which are in profit), 1 Guernsey Bull, 4 young heifers, 2 steers, and 13 pigs of various ages and 1 horse. We have been self-supporting for the last year, the Group Settlement Department only advancing us the manure for top-dressing of pastures. We only milked 7 cows last season, but they did very well and with the money obtained from the sale of pigs and bull service fees, we had a comfortable living and were also able to buy various things

for use on the farm. My wife is a great help to me, she thinks nothing of doing all the milking, separating, feeding calves and pigs herself, should I be away from home in the evening. We have three children, two girls and a boy six months old. Our eldest was the youngest baby when we left Leeds, she was then ten months old and is now eight years.[68]

Clearly, content to have a bigger farm on the original Leeds Group, Bean was doing well; but unforeseen, immense challenges lay ahead.

Leeds community cohesion under stress

Reconstruction inevitably undermined the coherence of the Leeds community in Northcliffe. Although there had been one incomer from another group, by 1929, only 12 of the original settlers remained. Dwindling numbers led to a decline in the school roll and, in February, it was closed. The building was dismantled and moved to a plot between Groups 142 and 143.[69] For the remaining Leeds settlers' children, school attendance now meant a journey of at least 8 miles. Absenteeism was common. Dissatisfaction among parents prompted letters and petitions to the Education Department in Perth. However, the director of education insisted that there were now too few pupils to justify the erection of another school closer to their farms. Instead, he suggested enrolment in "correspondence classes".[70] Parents disputed his assertion that there were too few children and requested that the cottage abandoned by the Jacobs family on their now-condemned block (9019) be made into a school.

In April 1931, a count in Groups 104 and 107 revealed that there were in fact 40 youngsters, of whom 22 were school age (i.e., 5–12 years old) and 13 were "under fives".[71] Education Inspector Victor Box visited in June and recommended that a dismantled cottage on the more central site of the original 107 encampment be re-erected as a school. Although the land at that site was swampy and waterlogged in winter, the Leeds parents agreed to "fix the place up" for that purpose.[72] On 9 November 1931, this makeshift school opened with Miss Sybil Weston as teacher; she boarded with George Moody's family.[73] Victor Box's choice certainly proved unwise; in one of several letters to the director of education, Frederick Bean complained about the building being "entirely surrounded by water in the winter months" and it being "extremely damp".[74] On the anniversary of its opening, George Moody organised a gathering attended by the Bishop of Bunbury. Writing to Birchall, Bishop Cecil repeated Moody's verdict that the school stood in "a miserable swamp place" and "all the children get sore throats", and added that he thought it "extraordinary that, with all the vacant land around, such a place was used".[75]

Broken promises regarding educational provision were matched by an ongoing failure to offer church ministry. Back in January 1925, the Archdeacon of Bunbury visited Leeds to appeal for help to fund the church in that diocese. Birchall asked the *Yorkshire Press* to appeal for donations to a "Leeds West Australia Group Church Fund". Frank Baker became treasurer and subscribers were asked to donate for five years. The target was to raise £240 annually towards the stipend of an Anglican minster serving Northcliffe and Pemberton. Efforts to promote the fund were redoubled in February when Birchall received a letter from Leeds settler Mary Jacobs, who told him the children did not have "their morning prayers" because the school master (i.e., Frank Smith) "don't believe in such rubbish".[76] Finally, in 1930, the Rev George Limbert (originally from Micklefield and educated in Leeds) was appointed priest for Northcliffe and Pemberton and travelled out to take up his post. Once there, he solicited Birchall for yet more money to buy a car; another appeal raised additional donations and funds were transmitted. Limbert's letters to Birchall (1931–1932) offer vivid insights into the prevailing economic circumstances and mood of Leeds families at a time of mounting financial hardship.[77] However, Limbert abruptly left to take up a post in a distant parish (taking the car with him), and Northcliffe's settlers depended on the ministry of Deaconess Mildred Margary, partly funded by England's Mothers' Union, and on Pemberton's lay reader, Mr Reynolds; both were nominally supervised by the rector, Rev W. Bushell, at Manjimup, more than 28 miles away.[78]

Although a church was built in Northcliffe during Limbert's time, both he and (later) Deaconess Mildred held services for the Leeds Group settlers in Moody's bungalow. As the local parish was unable to afford to pay its clergy, the Leeds Fund continued to send annual remittances until 1934. By then, however, according to Bishop Cecil, there was "scarcely anyone … living in Northcliffe" and in winter "in surroundings which have become so depressing, the deaconess [is] deployed elsewhere in the diocese". Due to financial pressures of the Great Depression, many subscribers to the Leeds Church Fund resigned, and its income collapsed. Moreover, at the end of 1933, Birchall and Baker were invited to support the northern branch of the "Bunbury Diocesan Home Association", which raised funds across England, and they decided to close the Leeds Fund and place its few remaining subscribers "under the wing" of that nationwide appeal. Birchall's efforts in Leeds by then, of course, had propped up church ministry in Northcliffe for more than eight years.[79]

The Great Depression and group settlement recapitalisation

Just as the Great Depression began to exert severe downward pressure on the value of agricultural products, the financial arrangements of group settlement in Western Australia were changed. Although settlers were originally assured that their capital debt would never exceed £1,000, the scheme's mounting losses prompted the State government to seek the agreement of Canberra and Westminster to a re-valuation (recapitalisation) of each holding (Crawford and Crawford 2003, p. 116). Soon after, each settler was made an individual client of the Agricultural Bank and became liable for the payment of all interest on all accumulated debts accrued to his holding (Gabbedy 1988b, p. 463). Farmers were required to sign mortgages with the bank, and many were shocked to find they owed in excess of £3,000.[80] Walter Grimshaw later blamed the chairman of the Agricultural Bank Commission for placing "fictitious valuations on many holdings" and, after deductions, many families were left with "no income at all".[81]

George Moody told Birchall in December 1932 that the "position here is really deplorable". He enclosed a table of his own monthly "cream returns" for the two previous years. Between January 1930 and December 1931, the price he received for butterfat fell by more than 60%, and Moody foresaw more reductions to come. "Had I paid my interest ... you can see ... I should have been broke".[82] However, those defaulting on their interest payments lived in constant fear of eviction.

A Royal Commission investigating the State's dairy farming acknowledged that few group settlers could meet interest payments "except at the expense of the development of the farm and herd, and the depletion of the bank's asset" (*Report of the Royal Commission on Dairy Farming in the South-West* 1932, p. 15). Walter Grimshaw reported that, by November 1933, only "Bean, Palmer, Bristow, Wilson, Downs, Moody and myself" were left of the original 19 Leeds settlers. However, Grimshaw also said that, personally, he was:

> ...at present carrying about 2½ tons of cream from this district, per week, into Northcliffe, for which I have a contract with the Co-operative Factory, so this will give you some idea of the general output. I pick up the cream at each farm, twenty suppliers, twice weekly. In spite of this though, the returns preclude any possibility of settlers meeting all their liabilities, chiefly on account of the abnormal valuation of our holdings.[83]

Debts were attached to the equity represented by the farm holdings, and for men unable to meet interest payments, still less pay down capital debts, a realistic option was to retire completely from the scheme. It was necessary speedily to find alternative work as the State disqualified them from receiving unemployment benefits (Crawford and Crawford 2003, pp. 116–7). There was considerable resentment when vacant farms were subsequently offered by the bank to others at much lower capital values.[84]

Moody wrote to Birchall to endorse the extraordinary mission to Britain of Leonetta Owen Tucker, wife of an Agricultural Bank inspector (based in Northcliffe) who arrived in November 1932 and remained until August 1933. She lobbied tirelessly on behalf of the "Settlers Association of Western Australia".[85] She delivered copies of the settlers' petition to all those she thought had influence, including Birchall, calling for a complete overhaul of the group settlement scheme. Somewhat unrealistically, the redoubtable

Mrs Tucker argued for intervention by the imperial government. She met Sir John Birchall several times and listened, from the Ladies' Gallery of the House of Commons, to a debate on Empire settlement.[86] She was fearless in her commitment to the group settlers' cause but, in fact, she returned to Australia with very little. Moreover, the authorities in Perth were angered by her activities and accused her husband, Arthur Tucker, of disloyalty and colluding in her campaign. The shabby denouement of the episode was Arthur's abrupt dismissal by the Agricultural Bank.[87]

Yet another Royal Commission affecting the group settlement scheme, this time focused on the Agricultural Bank, reported in October 1933. Amongst its recommendations, the report called for the "conditioning" of the bank's securities, which meant not only writing off accumulated interest debts on group settlement blocks but also reducing the principal sum deemed to be owed (*Report of the Royal Commission on the Agricultural Bank* 1933, p. 80). In due course, the capital debt of many remaining settlers was slashed by as much as two-thirds. By then, however, notwithstanding the report's recommended swift "liquidation of securities", that is, issuing mortgages on vacant farms to new occupiers, it was all too late to save hundreds of abandoned holdings; many had already reverted entirely to bush.

Soon after his appointment as Western Australia's Minister for Lands, and on completing an extensive tour of inspection of the group settlements, Frank Troy condemned the entire project. The State's taxpayers were meeting interest charges of £400,000 per year on the scheme's £10 million of accumulated debt. Of 6,000 original settlers, only 1,600 remained; indeed, Troy thought fewer than 300 were likely to succeed.[88]

In July 1935, Section 51 of the Agricultural Bank Act was invoked to allow mortgage repayments to be taken automatically from the proceeds of each farm. Northcliffe settlers found their monthly cream cheques plundered, leaving them little upon which to live. There was uproar, and eventually Northcliffe farmers carried out their threat to withhold the supply of cream to the butter factory.[89] Walter Grimshaw ignored the strike, continuing to collect cream cans as usual from fellow farmers unprepared to lose their income but, as a precaution, he made it known that he carried a gun (Gabbedy 1988b, p. 539). Sir John Birchall's feelings on learning the grim news of the settlers' plight can only be guessed.

Conclusion

In hindsight, the business model underpinning Western Australia's interwar group settlement was flawed. Moreover, the early promoters of the scheme distorted the facts regarding its potential for success (Brayshay and Selwood 2002, pp. 85–93).[90] Underestimates of the investment required and inadequate knowledge of the soils, topography and ecology of the forested areas earmarked for agriculture were also critical factors.

In Northcliffe, though the State's Forest Department protected consider-able areas of tree cover, the settlers' removal of so much woodland, previ-ously taking up huge quantities of water, inevitably altered soil hydrology and led to winter waterlogging and swamps. Any early advice regarding the low fertility and unsuitability of much of the area for sustainable pasture seems to have been ignored. Although the means to manage the soils and make them capable of yielding good pasture and other crops were eventu-ally devised, trials should have been conducted before large-scale settle-ment commenced. In any case, it is clear that many of the field supervisors and foremen lacked relevant skills and made serious errors when selecting where to focus the effort of clearing, with their choice being based on the relative ease of vegetation removal.

Farm block sizes (up to 160 acres) adopted in Western Australia's south-west group settlement scheme appear to have been modelled on the small family farms characteristic of nineteenth century Britain's "hay and dair-ying" districts; they were far too small to be viable in Western Australia's terrain. The so-called reconstruction of blocks between 1927 and 1929 was merely a tentative first step in recognising this fundamental mistake. In Northcliffe, by the end of the 1930s, much larger units that had a chance of profitability were at last being formed (Crawford and Crawford 2003, p. 125).

The idealised vision of clusters of closely spaced small farm ventures, pro-vided with rural community facilities for education, recreation, health care and worship was appealing, but the creation of that kind of human land-scape was a pipe dream. Moreover, the geographical remoteness of many of the Groups, including Northcliffe, precluded the diversification of farm enterprises to supply a mix of products: large concentrations of potential customers were simply too far away.

While attempts in the 1930s to recoup mounting government losses by mortgaging blocks to the Agricultural Bank and thereafter calculating an upward capital valuation made economic and political sense, coinciding with the damage inflicted by the Great Depression, the policy proved disas-trous. However, the resilience of the few remaining Leeds settlers is impres-sive. Many times, in letters to Birchall, they spoke of the magnificence and beauty of the landscape, but they also knew its potential for cruelty. On Wednesday 10 February 1937, a catastrophic bush fire swept through the entire region.[91] The farms on Group 107 were badly hit. No appreci-able rain had fallen since the previous October and, dry as dust, Walter Grimshaw's property was destroyed. John Downes lost his entire store of hay but was otherwise less badly affected, but the blocks of Frederick Bean and John Palmer were ravaged. Seventeen-year-old Clarence Palmer built a firebreak to save his father's paddocks, but when they saw huge flames advancing from another direction, they all ran for their lives. Clarence later returned to the Palmers' house and cowshed and filled cream cans at the creek to douse the walls, so saving both buildings. Elizabeth Grimshaw was "buried" for protection under sand by Walter, and afterwards said she

Figure 6.6 Jack and Laura Palmer on their Group 107 farm with their three daughters: Violet, Olive and Audrey, and Violet's baby daughter, circa 1935.

Courtesy: The Northcliffe Pioneer Museum, The Palmer Family Archives

"knew what hell was like". Laura Palmer wrote that "everybody round here and Northcliffe thought the world was coming to an end".[92]

Remarkably, in April 1938, Walter Grimshaw told Birchall that, with his son's help, he had "rebuilt everything". But, by then, of the original 19 settlers, only four remained: Grimshaw, Albert Bristow, John Downes and John Palmer (Figure 6.6). Those who had moved away were, according to Grimshaw, "scattered about throughout the state" though, in fact, a majority went ultimately to Perth.

Thus, in 1933, after five years spent on the Group 148 farm to which he had been moved when his Group 107 holding was condemned, Jim Jacobs finally ceased farming completely, sold up and moved first to Manjimup, and then to Perth. Joe Brearley gave up farming earlier still and, perhaps resuming his trade as a joiner, in 1930 he moved to Waroona, where he had a "comfortable living" but ultimately also settled in Perth.[93] Remarkably, from the original 1924 group of emigrants from Leeds, only Herbert Marriott and his family completely abandoned their new life in Western Australia and returned to Britain.[94]

As the woes of the Leeds settlers multiplied and numbers diminished, Sir John Birchall appears to have felt an acute sense of responsibility and regretted his original encouragement for the enterprise. He received a copy of the 1937 Dominions Office report on Western Australia's group settlement, which included a section on the Leeds settlers. The findings catalogued every negative event occurring since the settlers arrived in Northcliffe and offered a relentlessly depressing, but rather distorted, picture.[95] In any case, several of Birchall's settler correspondents repeatedly sought to reassure him that he was not to blame. The ever-articulate Grimshaw best summed up their sentiments:

> Do not have any regrets about having promoted the Leeds Group Settlement. Most of us are convinced that you were right. If there is any shame attached, it belongs only to the maladministration at this end. There is no doubt at all that the conception of the scheme was wonderful, but the administration rotten with corruption.[96]

Growth in Western Australia's south-western group settlement regions remained flat until after the Second World War, but then the 30-year so-called "long boom" began, which saw a steady revival in their fortunes. Although changing international circumstances and drastic shifts in demand, especially for dairy products, exerted an impact on the evolving character of agricultural enterprises, the processes of farm expansion, mechanisation and technical innovation enabled the resuscitation and more assured survival of farm businesses in the Northcliffe district. In addition to the regeneration of agricultural activity, the natural beauty and favourable climate of this special place in Australia have presented opportunities for tourism. Visitors are drawn by the lively interest in the group-settlement heritage, which adds much to the particular appeal of the Northcliffe area. For direct descendants of the Leeds settlers, shared pride in their Yorkshire forebears certainly persists.[97] The considerable areas of karri, marri and jarrah provide not only a basis for commercial forestry but are also valued as a recreational resource; the coast and the national parks of Warren, Shannon and D'Entrecasteaux are close by. Indeed, a portion of the sandy land that was once within the margins of Group 107 came inside the protected boundary of D'Entrecasteaux National Park when it was designated. The national park is now noted as a habitat of the rare "Queen of Sheba" orchid and, between September and November, promoted to walkers and tourists as a beautiful "wilderness zone" on the "Jarrahland wildflower trail".

In common with several of the former group settlement areas, since the 1970s, Northcliffe has also become part of the "post productive rural transition" whereby "alternative lifestylers" and early retirees have moved into the district to take up former holdings vacated by the group settlers and their families. In pursuit of a way of life that benefits from the beautiful surroundings, relaxed ambience and welcoming culture, sited well away from the mainstream, this new wave of migrants have inadvertently become agents of a wider revaluing of remote rural localities in Western Australia. In consequence, in more recent years, outside investors have seen opportunities to sell the "consumption of a commodified rurality" and local land and housing prices have sharply risen (Curry et al. 2001, pp. 109–24; Perry 2014; Jones et al. 2015, pp. 129–30). Once a place of hardship in the era of group settlement, Western Australia's South West is now viewed very differently as a zone of landscape amenity with considerable potential for prosperity.

Notes

1. Employment insurance excluded a significant proportion of the workforce. The published unemployment total of 2 million in 1922 is thus an underestimate (House of Commons, Debates, 10 April 1922, Vol. 122).
2. Concern is evident in the British Cabinet Unemployment Committee Minutes, The National Archives (TNA), CO 721/17, 15 October 1920.
3. The Dominions comprised the four independent, self-governing members of the Empire: Australia, Canada, New Zealand and South Africa.
4. Western Australia's soldier-settlement scheme was first aimed at returning servicemen of the Australian Imperial Forces (AIF), but British ex-soldiers were ultimately among the 5,200 taking up opportunities to settle land purchased, or earmarked, for farm development (Gabbedy, JP 1988a, pp. 31–47).
5. TNA, CAB 24/131, 29 September 1921. Oversea Settlement and Unemployment.
6. Empire Settlement Act, 1922 (12 and 13 Geo. 5, c. 13), His Majesty's Oversea Dominions. An Act to make better provision for furthering British settlement, 31 May 1922.
7. Mitchell arrived in London on 19 March 1922 and departed on 24 June.
8. "An empty land: Australia's call for men – room for millions" *The Times*, 22 March 1922, p. 8; "Men and money for Australia: Sir James Mitchell's offer" *The Times*, 12 April 1922, p. 7.
9. "Settlers for Western Australia: big scheme going ahead" *The Times*, 12 July 1922, p. 9
10. "Group settlement: a South-West experiment – twenty settlers wanted" *The West Australian,* 17 March 1921, p. 7.
11. Soon after, the programme was extended to the vast Peel Estate, located close to Perth.
12. State Record Office of Western Australia (SROWA), Lands Department, AN3/13/Acc. 1657/509/17, Box 67, Land for Returned Soldiers – Nornalup.
13. The ambitious target of 75,000 group settlers set by Mitchell in 1922 proved unrealistic.
14. Devon and Cornwall settlers were placed in Groups 113, 114 and 116. Catterick families (headed by ex-servicemen receiving prior vocational training in agriculture) went to Groups 126 and 127; the Scots settled Group 133.
15. Australian Broadcasting Corporation, Social History Unit: Talking History (Oral History), Joe Brearley [junior] interview 1982. In 1924, the Brearley family, comprising parents, Joseph and Alice, and their 15-year-old son Joe (i.e., the ABC's 1982 interviewee) and his brother and sister had arrived in Australia from Leeds.
16. Birchall was knighted in 1929 (Supplement, *The London Gazette*, 2 June 1929, 3666). He was a Leeds MP from 1918 until 1940 (House of Commons, Debates, 10 April 1922, Vol. 122).
17. As well as several hundred press cuttings and many other documents and publications, Birchall's archives contain more than 85 letters from 33 different correspondents. Among the letters, 27 were sent to the MP by ten of the original Leeds settlers, and they span the period 1924–1938.
18. House of Commons Debates, Hansard, 10 April 1922, http://hansard. millbanksystems.com/commons/1920s (accessed 16 July 2017). The Leeds Employment Exchange was among 21 in the UK then dealing with 10,000+ unemployed. In fact, that month, Leeds registered 16,110 "wholly unemployed" insured workers.
19. "Empire development: Australian premier's prediction – the population problem" *Yorkshire Post*, 13 December 1923. The private archives of Sir John Dearman Birchall (JDB Archives), held by Mark Birchall, contain all

these documents. Birchall retained a press-cutting agency (Durant's of Holborn) and so collected articles from both Yorkshire and national newspapers. Leeds settlers also sent him cuttings from Australia's newspapers.

20. Newcombe visited Western Australia between March and June 1923 (Brayshay and Selwood 1997a, pp. 206–8). Birchall's archive contains papers and press cuttings including reports of the sailing of the first Devon and Cornwall group, aboard the SS *Sophocles* ("Group Settlement", *The Times*, 1 February 1924, p. 12).
21. "A Leeds colony for Australia", *Yorkshire Evening Post*, 13 October 1923.
22. "Shifting Britain: group migration proposed scheme", *The Courier, Brisbane*, 17 October 1923.
23. JDB Archives, 31 October 1923, "One of your constituents" to Major Birchall. See also: "City colonists: can they rough it in a strange country? M.P.'s idea criticised" *Leeds Mercury*, 15 October 1932; "Leeds in Australia: personal impressions of bush life", *Yorkshire Evening Post*, 24 October 1923; "Leeds settlers for Australia: Empire migration – the problem viewed from both sides", *Yorkshire Evening Post*, 25 October 1923.
24. Among them, according to Birchall's notes, 213 expressed an interest in joining a "Leeds colony". He categorised their stated work experience and perhaps hoped to assemble a party of settlers with complementing skills. Such an aim proved impossible. The final approval of all applicants by Australia House was required.
25. "Leeds migration scheme: the first 120 families', *Yorkshire Post*, 27 October 1923.
26. JDB Archives, Lists of Applicants, 26 October 1923. Of the 19 men ultimately selected, and approved in London for the Leeds Group, only seven: Frederick Bean, Joseph Brearley, John Leese, Herbert Marriott, John Moody, Wilfred Smith and Herbert Wilson, were interviewed on that first evening. (The press reported the disappointment of "rejected" families, "who have not known work or a substantial weeks' wage for eighteen months"; see: "Leeds migration scheme: many disappointed applicants", *Yorkshire Post*, 3 November 1923.
27. JDB Archives, Great Northern Hotel, "Leeds in Australia" Dinner Menu, 26 October 1923. Birchall's 11 guests signed his menu card. Captain Frank Baker was to become one of his closest associates supporting the "Leeds colony".
28. Ibid., 4 December 1923, copy of the circular letter sent to the selected members of the Leeds Group.
29. Ibid., reports on visits made to families going to "New Leeds" 1924; 22 March 1924, Miss EO Lambourn to Major John Birchall. A few families needed winter coats, children's clothes and pairs of shoes, which were given. Some Leeds female settlers corresponded with the Yorkshire Ladies' Council after arriving in Australia. The press reported that, in contrast to the Devon and Cornwall group, some Leeds children were "indifferently clad" for a 12,000-mile journey (see: "The Leeds emigrants: suggestions for comfort of future parties', *Yorkshire Post*, 4 March 1924.
30. JDB Archives, 22 February 1924, LP Curguren to Major JD Birchall; Robert Scott (of Edward North's Book Saloon) to Birchall. Elizabeth Amm (née Moody) stated that her mother received *Jane Eyre*, and her father, *Moby Dick*, (Daubney 2001, p. 11).
31. For Christmas 1924, Major and Mrs Birchall sent each Group 107 family a calendar for 1925. In writing her thanks, Marriott's wife reported that the calendars had been printed by the company in Leeds of which she had earlier been an employee.
32. As a gift, Mrs Birchall appears to have presented each of the wives with a small evening bag; see: JDB Archives, 21 July 1924, Mrs H Marriott to Mrs [Adela] Birchall. ("... I shall always treasure my Bag you presented to us at the

Town Hall, in fact, if ever I have the privilege of meeting you again, I shall be pleased to show you it".)

33. "Leeds in Australia: civic God-speed to Empire settlers', *Yorkshire Post*, 25 February 1924.
34. JDB Archives, 25 February 1924, CA Hood to Major John Birchall.
35. Both Perth's Battye Library and Northcliffe's Pioneer Museum hold copies of the 28-page scrapbook deposited by Richard Grimshaw, Walter's son, containing his father's press cuttings.
36. JDB Archives, 2 October 1924, Mary Jane Jacobs to Major and Mrs Dearman Birchall.
37. With six children and another on the way, John and Laura Palmer had a shack to themselves.
38. JDB Archives, 31 January 1925, John T Leese to [Major] Birchall.
39. Ibid., 25 January 1926, J Brearley to Major Birchall.
40. Ibid., 18 April 1926, J Brearley to Major Birchall.
41. Ibid., 14 July 1924, WA Grimshaw to Major Birchall.
42. Fred Bean's wife, Edna, suffered permanent hearing impairment when she accidentally exploded a stick of gelignite as she made a bonfire of the broken logs from a tree that the menfolk were felling; see: Daubney, 2001, p. 21.
43. "20 months at 'Leeds in Australia': bringing down the big trees with gelignite", *Yorkshire Evening Post,*18 January 1926.
44. JDB Archives, 16 February 1925, W Grimshaw to Major Birchall.
45. Some of the blocks or holdings had better soils than others. In the ballot for the blocks on Group 107, Walter Grimshaw was lucky to be placed on the best land to be found anywhere in the Leeds settlement.
46. JDB Archives, 7 July 1934, J Unsworth to Sir John Birchall, enclosing the British Ex-Service Legion of Australia's "memorial" of complaint on behalf of the group settlers. See also SROWA, AN 3/3 Acc. 541, Item 7529/23 Group Settlement – Drainage of the Northcliffe Area, 28 July 1924, 1 August 1924, fos 22, 23.
47. JDB Archives, 16 June 1929, GF Moody to Sir John Birchall.
48. Ibid., 31 January 1925, John T Leese to Major John Birchall.
49. SROWA, Education Department. Acc. 1497, Item 1745, Groups 104 and 107, Building Works, fo. 54.
50. Ward was another Yorkshireman, previously practising in Morley, five miles from Leeds.
51. JDB Archives, 16 February 1925, Walter Grimshaw to Major John Birchall. Unfortunately, two bottles in the chest had been broken in transit, damaging some of the drugs.
52. Ibid., 2 October 1925, Mary Jane Jacobs to Major John Birchall. Through Birchall's good offices, Mrs Jacobs was able to nominate her daughter's family (the Doyles), and they joined the Group Settlements in due course. (Note: Mary was a widow with a daughter when she married James Jacobs. By 1926, Mary and James had two more daughters.)
53. One of Herbert Bayliss's sons sustained serious injuries in an accident, and John Smith's departure involved a family matter.
54. JDB Archives, 21 July 1924; 29 January 1925, Mrs H Marriott to Mrs Birchall.
55. "Back from Australia, Leeds engineer and the settlement scheme: hopeless", *Yorkshire Evening Post*, 23 June 1927; "Leeds in Australia: experience of a returned settler', *Yorkshire Evening Post*, 5 August 1927.
56. JDB Archives, 9 June 1926, HP Colebatch to Major John Birchall; 11 July 1927, Herbert Marriott to Major John Birchall (the Marriott family were by then living at 40 Privilege Street, Wortley, Leeds); 10 August 1927, WC Angwin to Major JD Birchall.

57. Ibid., 24 September 1927, Walter Grimshaw to Major John Birchall.
58. Ibid., 18 April 1926, Joseph Brearley to Major John Birchall.
59. Ibid., 16 June 1926, George Moody to Sir John Birchall. Moody reported a case of a repossessed horse on Group 106. The horse's confiscation for many weeks deprived the settler of his draught animal and means of transport, but the cost of its feed was nonetheless added to the debt attached to the block.
60. "Making the rough places plain: road construction in the new Leeds", *Yorkshire Evening Post*, 29 January 1926. Three of the "road gang" boarded with Mrs and Mrs Herbert (Bert) Wilson.
61. See also Latham, CG 1928, "Prospects of land settlement and farming in Western Australia in: *Report of Proceedings of a Meeting of the Empire Parliamentary Association on Migration and Land Settlement within the Empire, 4 July 1928*, Empire Parliamentary Association, London, pp. 3–23; Philip Collier had replaced Sir James Mitchell as the State's premier on 17 April 1924.
62. George and Annie Johnson vacated their farm on 23 March 1926 and moved to Perth. Edgar Foxcroft also moved to the city, and the two families remained close friends for several years thereafter.
63. JDB Archives, Mildred R Ormerod to Major Birchall, February 1929. Mrs Ormerod enclosed the letter from Mrs J Leese.
64. Ibid., 28 February, 11 June and 27 June 1929, WC Angwin to Major John Birchall. In his third letter, Angwin told Birchall of John Leese's further move to Group 2.
65. There were also other reasons for Twigg's dismissal from Group 107.
66. JDB Archives, Mildred R Ormerod to Major Birchall, 29 June 1929.
67. Ibid., GF Moody to Major Birchall, 16 June 1929; WC Angwin to Sir John D Birchall, 23 July 1929, 22 October 1929; ASC Chatley to J Fox, 2 September 1929.
68. JDB Archives, F Bean to Sir John Birchall (May?) 1931. Sir John noted that he wrote a "long letter" in reply on 31 July 1931. Bean's optimism crumbled soon thereafter, as collapsing agricultural commodity prices in the Great Depression began to bite, and he eventually gave up and left. Seven years earlier, Dorothy Bean, then ten months old, was photographed with her parents, Frederick and Edna and featured in: "Leeds migrants steaming for the sunny southern seas: the youngest emigrant", *Leeds Mercury*, 3 March 1924.
69. SROWA, Education Department. Acc. 1497, Item 1745, Groups 104 and 107, Building and Works, fos 101, 104.
70. Ibid., 20 February 1930, fo 4.
71. Ibid., Education Department. Acc. 1497, Item 285/32 Group Settlement: Groups 104 and 107, Building and Works, fos 9–10.
72. Ibid., Item 1745, Groups 104 and 107, Building and Works, fos 10; 28 June 1931, Victor Box to Director of Education, fos 20–23.
73. JDB Archives, GF Moody to Sir John Birchall, December 1932. SROWA, Education Department. Acc. 1497, Item 1745, Groups 104 and 107, Building and Works, fo 105.
74. SROWA, Education Department. Acc. 1497, Item 283/32, Building and Works, 8 February 1932, F Bean to Director of Education, fos 57–58; 29 March 1934, note recording closure of Group 104 and 107 school "because of poor attendance", fo 93.
75. JDB Archives, 25 November 1932, Bishop of Bunbury (Cecil Wilson) to Sir John Birchall.
76. Ibid., 11 February 1926, Mary Jane Jacobs to Major Birchall.
77. Ibid., 22 December 1930, 26 February 1931, 2 May 1931, 23 July 1931, GW Limbert to Sir John Birchall; 1931 [no precise date] Rev. GW Limbert to the Society for the Propagation of the Gospel, London (copy); 31 December 1931, GW Limbert to FS Baker (copy).

78. Ibid., 18 June 1934, Bishop of Bunbury to Frank Baker (copy).
79. Ibid., 28 January 1926, Birchall and Baker, call for church fund subscriptions; 24 July 1934, FS Baker to Sir John Birchall; September 1934, FS Baker and JD Birchall to subscribers to Leeds West Australia group church fund.
80. Ibid., July 1934, British Ex-Service Legion of Australia, Western Division, Perth, memorial of complaint on behalf of group settlers.
81. Ibid., 8 April 1938, W Grimshaw to Sir John Birchall.
82. Ibid., December 1932, GF Moody to Sir John Birchall.
83. Ibid., 8 November 1933, W Grimshaw to Sir John Birchall.
84. Ibid., 7 July 1934, J Unsworth to Sir John Birchall.
85. Ibid., 11 October 1932, GF Moody to Sir John Birchall; 24 November 1932, Leonetta Owen Tucker to Sir John Birchall.
86. Ibid., 30 July 1933, Leonetta Owen Tucker to Sir John Birchall.
87. Ibid., 2 February 1934, 14 June 1934, Leonetta Owen Tucker to Sir John Birchall. According to the cemetery records, Mrs Owen Tucker, born in 1892, died in Margaret River, aged 90, in 1982. She was therefore in her early 40s when she engaged in her extraordinary campaign on behalf of the group settlers. Her children, Eustace and Phyllis, were apparently cared for by their father, Arthur Tucker, while Leonetta travelled to England.
88. "Group Settlement in Australia: unsatisfactory conditions', *The Times*, 1 December 1933, p. 13.
89. "Aggrieved settlers: growing tension – appeal to Lieut.-Governor", *The West Australian*, 3 August 1936, p. 19.
90. Ibid., July 1934, The British Ex-Service Legion of Australia's "memorial".
91. "Northcliffe blaze", *The West Australian*, 13 February 1937, p. 20; "Bush fires and McCallum: Northcliffe's last legs – settlers' terrible privations", *The Workers Star*, 26 February 1937, pp. 1 and 3.
92. Northcliffe Pioneer Museum, Palmer Family Archives, 21 February 1937, Laura Palmer to Violet Palmer.
93. JDB Archives, 8 April 1938, W Grimshaw to Sir John Birchall. Detailed evidence of families after they left their group farms is somewhat patchy; however, it has been possible, in several cases, to reconstruct the lives of settlers' children. For example, Elizabeth Lucy Moody, born in Harrogate on 6 April 1920, was exactly four years old when she disembarked from the *SS Diogenes* at Albany. She had travelled with her parents from Middleton in Leeds to Northcliffe's Group 107. As an adult, Elizabeth worked in Perth as a cook. She died, aged 82 years, on 20 January 2003.
94. Northcliffe Pioneer Museum, Palmer Family Archives, 21 February 1937, Laura Palmer to Violet Palmer; JDB Archives, 8 April 1938, W Grimshaw to Sir John Birchall.
95. TNA, DO (Dominions Office) 35, 686/6, Western Australia Group Settlement, Leeds Group Report, 1937.
96. JDB Archives, 16 November 1935, W Grimshaw to Sir John Birchall.
97. Northcliffe Pioneer Museum, "'All of them came from Yorkshire': descendants of the Northcliffe District's Group 107 pioneers who are currently living in Australia", 2012, updated 2014, pp. 1–18.

References

Amery, LS 1953, *My Political Life, Volume Two: War and Peace, 1914–1929*, Hutchinson, London.
Bankes Amery, W 1926, *Report on the Group Settlements in Western Australia*, HMSO, London.

Blond, PEM 1987, *A Tribute to Group Settlers*, University of Western Australia Press, Nedlands.

Bolton, G 1994, *A Fine Country to Starve in*, rev. ed., University of Western Australia Press, Nedlands.

Brayshay, M and Selwood, J 1997a, "Devon and Cornwall group settlement in Western Australia in the interwar years", *Transactions of the Devonshire Association for the Advancement of Science, Literature and the Arts*, Vol. 129, pp. 199–248.

Brayshay, M and Selwood, J 1997b, "The Devon and Cornwall Group Settlements of Western Australia" in: E Bliss (ed.), *Island: Economy, Society and Environment*, New Zealand Geographical Society Conference Series, No. 19, Palmerston North, New Zealand, pp. 344–8.

Brayshay, M and Selwood, J 2002, "Dreams, propaganda and harsh realities: landscapes of group settlement in the forest districts of Western Australia in the 1920s", *Landscape Research*, Vol. 27, No. 1, pp. 81–101.

Constantine, S 1980, *Unemployment in Britain between the Wars*, Longman, Harlow.

Constantine, S 1984, *The Making of British Colonial Development Policy, 1914–1940*, Frank Cass, London.

Crawford, P and Crawford, I 2003, *Contested Country: A History of the Northcliffe Area of Western Australia*, University of Western Australia Press, Crawley.

Curry, GN, Koczberski, G and Selwood, J 2001, "Cashing out, cashing in: rural change on the south coast of Western Australia", *Australian Geographer*, Vol. 32, No. 1, pp. 109–24.

Daubney, A 2001 (Ed.) *Northcliffe: "I Remember When", Fourteen Families Remember*, Alison Daubney, Northcliffe, Western Australia.

Devon and Cornwall Migration Committee 1923, *Western Australian Government Group Migration Scheme with the Co-Operation of Devon and Cornwall*, Three Towns Printing, Plymouth.

Drummond, I M 1974, *Imperial Economic Policy, 1919–1939*, University of Toronto Press, Toronto.

Fedorowich, K 1990, "The assisted emigration of British ex-servicemen to the dominions, 1914–1922", in: S Constantine (Ed.) *Emigrants and Empire: Settlement in the Dominions between the Wars*, Manchester University Press, Manchester, pp. 45–71.

Gabbedy, JP 1988a, *Group Settlement, Part 1: Its Origins: Politics and Administration*, University of Western Australia Press, Nedlands.

Gabbedy, J P 1988b, *Group Settlement, Part 2: Its People: Their Life and Times – An Inside View*, University of Western Australia Press, Nedlands.

Group Settlement in Western Australia at Work on the Peel Estate: Reclaiming a Great New Province 1923, Government of Western Australia, Government Printer, Perth.

Hunt, IL 1958, "Group settlement in Western Australia: a criticism", *Studies in Western Australian History*, Vol. 3, pp. 5–42.

Johnson, M 2005, "'Promises and pineapples': post-First World War soldier settlement at Beerburrum, Queensland, 1916–1929", *Australian Journal of Politics and History*, Vol. 51, No. 4, pp. 496–512.

Jones, R, Diniz, A, Selwood, HJ, Brayshay, M and Lacerda, E 2015, "Rural settlement schemes in the south west of Western Australia and Roraima State, Brazil: unsustainable rural systems?", *Carpathian Journal of Earth and Environmental Sciences*, Vol. 10, No. 3, pp. 125–32.

Jupp, J 1990, "Immigration: some recent perspectives", *Australian Historical Studies*, Vol. 24, No. 4, pp. 285–91.

Kennedy, D 1988, "Empire migration in post-war reconstruction: the role of the Oversea Settlement Committee, 1919–1922", *Albion*, Vol. 20, No. 3, pp. 403–19.

Langfield, M 1999, "Recruiting immigrants: the First World War and Australian immigration", *Journal of Australian Studies*, Vol. 60, pp. 55–65.

Lillis, JB 1926a, "New Leeds in Australia", No. 1, *Yorkshire Evening Post*, 11 March 1926, p. 6.

Lillis, JB 1926b, "New Leeds as it is today, No. 2: Evening Post investigator chats with settlers in Australia", *Yorkshire Evening Post*, 12 March 1926, p. 8.

Lillis, JB 1926c, "New Leeds as it is today, life of settlers in Australia, No. 3: Yorkshire Evening Post investigation", *Yorkshire Evening Post*, 13 March 1926, p. 12.

Morley, F 1922, "The incidence of unemployment by age and sex", *The Economic Journal*, Vol. 32, pp. 477–88.

Morrison, J C 1923, *A Unique Experiment: Farms and Farmers in the Making. Western Australia's Groups*, Western Australia Government Printer, Perth.

Oversea Settlement Committee 1923, *Report for the Year Ended 31 December 1923*, HMSO, London.

Perry, C 2012, *Dr Lionel Frederick West: 1868 to 1929: Northcliffe's Only Doctor*, Helvetica Press, High Wycombe, Western Australia.

Perry, C 2014, *Northcliffe: The Town That Refused to Die*, Digger, Albany, Western Australia.

Perry, M 2000, *Bread and Work: Social Policy and the Experience of Unemployment, 1918–39*, Pluto Press, London.

Pope, D 1977, "The contribution of United Kingdom migrants to Australia's population, employment and economic growth: federation to the depression", *Australian Economic Papers*, Vol. 16, pp. 194–210.

Powell, JM 1981, "The debt of honour: soldier settlement in the dominions, 1915–40", *Journal of Australian Studies*, Vol. 8, 64–87.

Report of the Royal Commission on Group Settlement 1925 Western Australian Government Gazette [WAGG] No. 48, Western Australian Government Printer, Perth.

Report of the Royal Commission on Dairy Farming in the South-West 1932, WAGG No. 8, Western Australian Government Printer, Perth.

Report of the Royal Commission on the Agricultural Bank 1933, WAGG No. 47, Western Australian Government Printer, Perth.

Roe, M 1990, "'We can die just as easy out here': Australia and British migration" in: S Constantine (Ed.) *Emigrants and Empire: Settlement in the Dominions between the Wars*, Manchester University Press, Manchester, pp. 96–120.

Roe, M 1995, *Australia, Britain and Migration, 1915–1940: A study of desperate hopes*, Cambridge University Press, Cambridge.

Selwood, J and Brayshay, M 1998, "From one room to another in the great house of the Empire: Devon and Cornwall group settlement in Western Australia", *Journal of the Royal Western Australian Historical Society*, Vol. 11, pp. 476–95.

Shann, E 1925, "Group settlement of migrants in Western Australia", *Economic Record*, Vol. 1, No. 1, pp. 73–93.

Williams, K 1990, "'A way out of our troubles': the politics of Empire settlement" in: S Constantine (Ed.) *Emigrants and Empire: Settlement in the Dominions between the Wars*, Manchester University Press, Manchester, pp. 22–43.

7   The Hudson's Bay Company's sponsored agricultural settlement at Vermilion, Alberta, Canada

John Selwood
James Richtik

Introduction

In the 1920s, the Hudson's Bay Company (HBC) found itself still with a massive land inventory in the Canadian prairies after surrendering its monopoly trading rights in Rupert's Land in 1870 and successfully negotiating ownership of extensive landholdings in return (Galbraith 1951). This land was distributed fairly evenly over the prairies, with the company being granted nearly two sections (square miles) of land in virtually every 36-square-mile township laid out in the region (see Figure 7.1). Although the HBC had sold or otherwise disposed of much of this land by the 1920s, it still owned several million acres and was eager to avoid increasingly burdensome land taxes and to turn the land into revenue for its shareholders or to finance its other endeavours. Unfortunately for the HBC and the settlers it recruited for the settlement scheme, they were put on the land at the onset of worldwide economic calamity and a disastrous period in the history of western Canadian agriculture. The 1930s in western Canada, dubbed the "dirty thirties", produced many tales of woe from farmers caught in the combination of poor yields and low grain prices that prevailed over the decade. Most of these tales concerned individuals struggling on their own and described the failure of many of them in the face of natural and economic adversity. However, as this chapter will show, even the financial support of government and private enterprise did not ensure success. Despite its reliance on settlers from Britain, and preferably from its provinces, the HBC Scheme did little to support the notion that country of origin would have a positive effect on their success rates (Nichols 1928).

The HBC's interest in taking advantage of the Empire Overseas Settlement Scheme first manifested itself in the creation of a subsidiary company, the Hudson's Bay Company Overseas Settlement Limited (HBCOSL) in September 1925. This company initiated various programs to identify opportunities for placing migrants in the service of the HBC and to provide labour for Canadian farms already in operation. The HBCOSL invited "enquiries from families, widows, single women, single

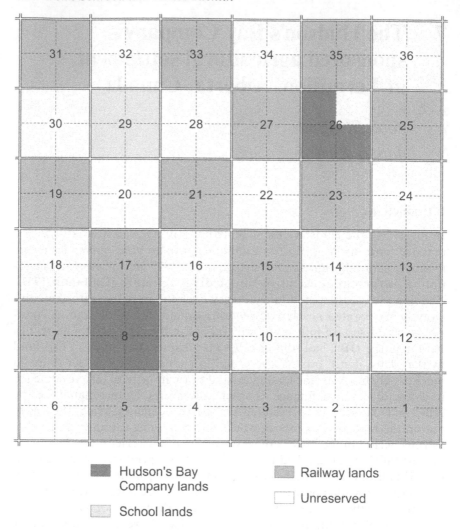

Figure 7.1 Land allocation within a typical prairie township.

Source: Tyman JL 1972, *By Section, Township and Range*, Brandon University, Brandon, Manitoba, p. 22

men and youths desiring to purchase and settle on Canadian land, or to work for farmers in Canada" (HBCOSL 1926). The HBCOSL even established a farm training programme at Brogborough Park Farm at Ridgmont in Bedfordshire to help potential migrants qualify for jobs in Canada by making them "semi-experienced" and thereby eligible for the "cheap passage rate" (HBCOSL 1928). The HBC also sought out government and corporate partners for its plans to settle farmers on company landholdings.

In 1926, the HBC wrote to the Canadian Minister of Immigration and Colonization that "the time is ripe ... for opening wide the gates of opportunity" and that the HBC would be willing to make available 2,000 farms and the Canadian Pacific Railway (CPR) another 1,500 for just such a purpose considering "your Department has now practically exhausted the improved farms owned by the Government available for colonization under the Three Thousand Family Scheme", previously used to attract British immigrants. Under the proposal, Ottawa would supervise land selection, the British Government would fund setting up farms to a maximum of $2,000 per farm, the CPR would oversee the setting up of the farms and aid the settlers and the HBC would assist in selection of settlers. The farms were to be on land owned by the CPR and the HBC. The letter also indicated that 1,000 farms could be prepared in 1927, another 1,000 in 1928 and 1,500 in 1929. In each case, families would be sent out the year after preparations had been made (Sale and Beatty 1926). Nothing came of the proposal in the immediate term, but it became a model for the scheme eventually adopted.

In May 1927, the HBCOSL formed a partnership with the Cunard Steam Ship Co. Ltd. to undertake a more modest version of the earlier proposal. Part of the motivation was to "prove that the government's "3,000 Family Scheme" was unnecessarily expensive, and that equally satisfactory settlement arrangements can be made by private Companies and Societies at much less cost" (Howard 1929). The CPR was persuaded to join the partnership in January 1928, and the three companies sponsored the farm migration scheme at Vermilion, which was eventually pieced together in mid-1928 in collaboration with the British Overseas Settlement Committee and the Canadian government. It called for "carefully selected British families [to] be placed on farms already prepared for them" (HBCOSL 1928), although by this time the number of farms had been reduced to 200. Half of these were to be located on HBC land and half were to be located on CPR land. The agreement as signed required that each family consist of a man and wife, with at least one child and a minimum of £50 on hand. They also had to be eligible for assisted passage under the agreement between the British and Canadian governments. For its hundred farmers, the HBC undertook to supply each family with a farm of 160 acres, a house, five acres broken and fenced and buildings and improvements not to exceed $1,200 in cost; to provide assistance for the families and find jobs for them if needed during the first year; to undertake collection of payments and not to charge more than "scheduled prices" for the land. The British Secretary of State agreed to advance up to $800 per family plus 50% of costs, excluding land – the total costs for each family not to exceed $5,000 (HBCOSL 1928). The agreement was changed almost at once. Because not enough families with £50 pounds on hand could be found, the British government put up another $200 per family and the HBC a further $50. Furthermore, because the CPR was

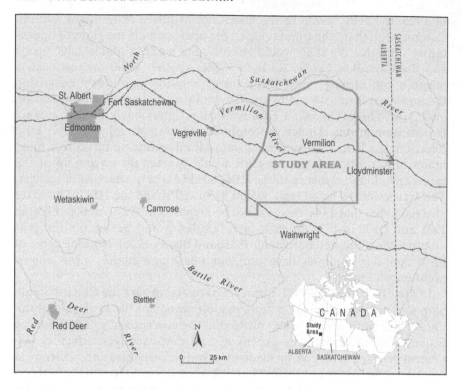

Figure 7.2 Location of the Hudson's Bay Company farm settlement scheme.

Source: Hudson's Bay Company Archives

allocating 320-acre farms for most settlers, the maximum costs allowable were raised to $7,000 for the larger farms (Brooks 1929; HBC 1929). This adjustment affected very few of the HBC's farms, most of which were only 160 acres in area and cost between $12 and $15 an acre. Money was also advanced for fire insurance to protect the initial investment. The arrangement was that the settlers would sail on Cunard or CPR ships and would be transported across Canada by the CPR. All three groups would participate in the recruitment of settlers and would share the scheme's administration costs (Chester 1929; Whishaw 1929).

Three possible locations for the HBC portion of the scheme were considered – two in Saskatchewan and the one that was finally selected, which lay between Clandonald and Vermilion, in Alberta (see Figure 7.2).

This area was chosen largely because the CPR had decided to place some of its new settlers in the Clandonald area adjacent to its 1926 settlement scheme and because there were other British settlers already located nearby. The HBC's internal instructions to its Lands Department called for the selection of "one hundred good quarter-sections in British communities and as contiguous as possible" (Harman 1928). They were also to be located as close as possible to the CPR selections. Accordingly, the HBC's land in eight townships was removed from general sale at once (HBC 1928; Harman 1928; 1928a; Harmon 1928b; Garnett 1937). Other land was added later as needed, most of it between the Canadian National Railway's (CNR's) Vermilion and Wainwright lines where the CPR had chartered a branch line to be extended west from the railhead at Paradise Valley (Dennis 1928). The company later claimed it chose the district "because it has the reputation of being one of the coming districts in the province, the land being what is known in Canada as "park land", a type of land for the most part clear and ready for settlement, but with some poplar and willow bushes that could be used for erecting small buildings, fencing and firewood" (HBC 1930). The district in question was mostly what the Canadian Land Inventory now classifies as suitable for cropping, and the properties selected were at least as good as the average for the area. In addition, the locality had such a good reputation of being hail-free that the HBC encouraged not taking out hail insurance (HBCOSL 1930; CLI 1970). As it transpired, this was ill advised because hail was to cause havoc soon after the settlers began working their farms. The distribution of farm properties selected for the scheme is shown in Figure 7.3. The CPR allocated 37 farms in the general vicinity of Vermilion but centred on Clandonald. In meeting the rest of its commitment, the CPR selected properties at a more distant location at Wood Mountain in Saskatchewan (Whishaw 1929).

The HBC's first selections were located close to the CPR's farms at Clandonald, but these were limited in number because some of the land further to the west was already occupied by Ukrainians; and the HBC had promised its recruits they would be placed on farms near other British settlers (Harman 1928; Lupul 1988, p. 32). Even so, EH Gamble, the HBCOSL's local manager, subsequently appealed to senior management to have additional British settlers assigned to the district so as to "stiffen up" the area with a stronger British presence (Howard 1929). The next group of farms was located near Vermilion, which had been established as the administrative centre for the scheme, while later selections were further south. The HBC lands not used south of Vermilion were mostly marginal or submarginal for crops because of sandy soils and rough topography.

In preparing for the settlers' arrival, the HBC and CPR worked in close cooperation. In fact, the HBC had wanted a joint venture with the CPR because of the latter's experience with such settlement schemes. The CPR and the HBC used their separate sources for the supply of materials and, after comparing prices, generally used the cheapest suppliers for livestock, implements and other equipment. The CPR planned to use the same contractor they had previously used for building, but the HBC suggested it could beat the price. The HBC also wished to complete well drilling in advance of house construction to ensure that water would be handy (HBC 1930). Thus, although the settlers were only modestly provided for, they were nevertheless supplied with the basic necessities.

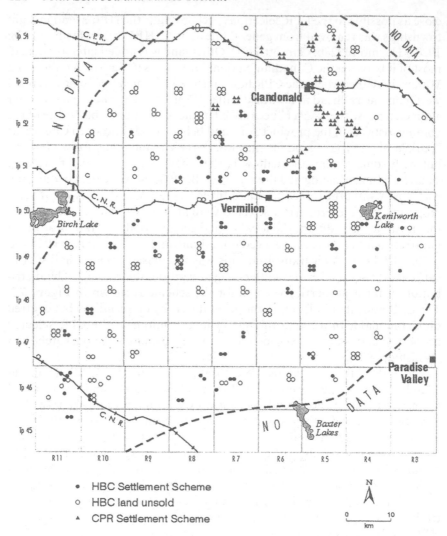

Figure 7.3 Location of the Hudson's Bay Company and Canadian Pacific Railway farms at Vermilion.

Source: Hudson's Bay Company Archives

In actuality, the tidy, preplanned arrangements for recruitment, selection, passage and placement of the settlers became a lot more muddled. After considerable delay, the HBCOSL received final clearance from officials in Ottawa in March 1929 for the placement of the 200 families on the farms, and selection of the migrants could proceed (Howard n.d.). The more experienced officers of the CPR and Cunard did most of the actual recruitment, with the HBCOSL London office being "merely a clearing house" (Howard 1929, 1929a). However, EF Bowkett, the HBCOSL's man responsible for the final selection of families for the company's holdings, vetted the applicants (Howard 1929b). Most of the scheme's recruits were successfully marshalled for passage by mid-April, but although several of them arrived in Canada together on the Cunard line's SS *Andania*, the remainder followed on in dribs and drabs, sailing on a number of different ships, landing at different ports and travelling out west on the CPR and the CNR (HBC 1931). The migrants were divided almost equally between Cunard's ships (*Andania*, *Ansonia*, *Ascania*) and the CPR's "Duchess" class ships. Although most docked at Montreal, others sailed into Quebec or even Halifax, which was not served by the CPR. Arriving in Halifax created difficulties because the migrants had to travel west on the CNR for at least part of the journey. The CPR's insistence that they change over to the CPR at the earliest opportunity was frustrating for the HBCOSL because most of the settlement was better served by the CNR. These logistical problems were overcome with only moderate expense, but the delay of arrival made things more difficult for the settlers (Peirson 1932).

The farms for the settlers had been prepared in 1928 and early 1929 in anticipation of their arrival. True to the contractual agreement, the settlers were placed on farms with the land partly broken and with basic accommodations supplied (see Figure 7.4). They were also issued with necessary farm

Figure 7.4 Sketch of cottage as provided on farm.

Source: Hudson's Bay Company Archives

implements, tools, vegetable and seed grain, and livestock sufficient to get them started. For the majority, these items became a charge against the settlers' accounts. Unfortunately, the first settlers did not arrive until mid-1929, and most were placed on their farm too late to plant crops. Worse yet, although 1928 had been an excellent crop year, 1929 was not, and those who did manage to get in a crop had poor yields because of a severe drought. Furthermore, due to the dry year, the land was so hard that breaking it was difficult, and the settlers prepared only half as much for the following year as had been expected. In addition, the poor yields pushed up local food prices, and there was little harvest work available in the district. As a result, most newcomers needed additional advances to make ends meet. Of the 95 families placed in 1929, nine had moved off their farms within the year. Another two families were deported back to England because they had been in the country for less than five years and were receiving municipal relief (HBC 1930).

In 1930, prospects appeared much better. Excellent weather produced the "finest and most regular crops that had ever been seen" (HBCA 1931). Wheat promised more than 30 bushels per acre and oats 77 bushels, and so more money was advanced to buy binders and twine. The settlers had averaged 30 acres cultivated in the previous year but had increased this to 40 acres in 1930. However, a wet fall delayed harvest and only 10% had been threshed when a blizzard hit in October. As a result, yields were reduced by 50%, and grain quality fell from No. 1 grade to Nos. 2 or 3. Wheat prices, already low in September, tumbled further over the winter so most wheat sold for only 30 cents a bushel. Nonetheless, more than half the settlers were expected to be able to make a partial payment on their farms. Although many families were in dire financial straits and asking for more advances, only 45 were granted them – for a meagre total of $2,087. By the end of the year, four more families had been deported. However, 14 new families were brought in as replacements, bringing the numbers on hand back up to 99 (HBC 1931).

Despite the initial setbacks, the HBCSOL continued in its efforts to recruit new settlers. Disappointed at the efforts of their partners (who were busily recruiting on their own behalf) the company used its London office for recruitment and propagandised the settlement scheme in its widely circulated in-house magazine, *The Beaver*. Advertisements showed idyllic scenes of a farm family busying themselves around their log cabin, nestled amongst the trees (Figure 7.5). Announcements heralded the rapid progress of the settlers already in place on their farms at Vermilion and listed the results of the company's "Farm Competition Awards". These were at pains to point out the diverse backgrounds and past occupations of the winners (see Figure 7.6). English, Irish, Scots and Welsh were identified. A minority of the winners had farm backgrounds, with the remainder being drawn from such varied backgrounds as mining and the transportation industry (HBC 1930a).

Figure 7.5 Cottage image as appearing in advertisement.

Source: Hudson's Bay Company Archives

Unfortunately, these achievements were not an accurate forecast of the settlers' future. From 1931 to 1937, the area experienced a number of dry years separated by years cursed by either frost or hail, or both. Yields were low everywhere but especially for the HBCOSL settlers, who usually had lower than the average yields for the district. For example, even in 1938, the district average was between 15 to 19 bushels an acre, whereas the HBCOSL farmers averaged only 13 bushels an acre (HBC 1938). From 1938 to 1940, there were good crops but, with few exceptions, poor crops were widespread throughout the 1940s (HBC 1951). Needless to say, this discouraged the settlers, driving many of them off their farms, despite the HBC periodically being moved to support the struggling farmers with various forms of assistance. The company generally provided only modest advances, feeling that "coddling the settlers would make them less self-reliant" (Brooks 1930). Company officials claimed that some farmers deliberately did poorly and refused to accept the company's help so as to qualify for relief and thereby win a return trip to England (HBC 1932). However, few settlers actually did this, and many battled on as the company reduced the price of their farms, rescheduled debts and waived interest payments. Notwithstanding these inducements, the settlers were constantly behind in repayment of their loans, leading company officials to insist that scheme members "could pay if they had the will to pay" (Stuart 1949). Meanwhile, the attrition continued. In 1934, the HBCOSL began to foreclose on farms where there was

Hudson's Bay Company
British Family Settlement
Vermilion, Alberta

FARM COMPETITION AWARDS

To encourage the progressive development of their new homes and farms, their gardens, and increase in cultivated lands and livestock, a number of prizes were offered for general competition.

The high standard of progress attained by so many of the farmers in close competition made it necessary to award fourteen consolation prizes of Hudson's Bay Company white woollen blankets.

The list of competitions and awards are as follows:

ALL ROUND BEST FARM—
First Prize (Cow)—Won by B. Griffiths, Farm No. 26, previously a Welsh miner.
Second Prize (Paint for Buildings)—W. Brown, Farm No. 69, previously a Northumberland motor haulier.
Consolation—A. C. Gillies, Farm No. 47, a Scotch farm manager.

LAND DEVELOPMENT—
First Prize (Hay Mower)— R. Gilholme, Farm No. 32, previously an English farm worker.
Consolation—B. Conlon, Farm No. 28, an Irish labourer; D. Geddes, Farm No. 40, a Lancashire market gardener; A. C. Monro, Farm No. 41, a Scotch farm worker.

LIVESTOCK—
First Prize (Brood Sow)— E. Foster, Farm No. 60, a farmer from Ireland.
Consolation—J. C. Brown, Farm No. 39, an English farm worker, etc.

HOUSE AND GARDEN—
First Prize (Churn)—E. J. Lugg, Farm No. 59, an English ship's painter from Cardiff, Wales.
Second Prize—R. Allison, Farm No. 51, a Yorkshire farm worker.
Consolation—E. C. Hughes, Farm No. 9, a Welsh miner; T. Payne, Farm No. 6, an Irish miner.

DAIRYING—
First Prize (a Cream Separator donated by Massey-Harris Company)—Wife of A. F. Holland, Farm No. 25, a licensed victualler and miner from Wales.
Consolation—Wife of D. Williams, Farm No. 22, a Welsh miner; wife of J. Little, Farm No. 82, English motor bus driver; wife of H. R. Evans, Farm No. 57, from Ceylon tea plantations.

CHILDREN'S ESSAYS—
First Prize—Moira Foster. Second Prize—Catherine Geddes. Third Prize—Albert Teasdale.

OTHER CONSOLATION AWARDS—
W. H. Hollyoake, Farm No. 30, an English labourer. R. M. Mathewson, Farm No. 71, a Scotch farm worker. W. G. Slee, Farm No. 58, a railway guard from Wales. A. H. Hartley, Farm No. 8, an English farm worker.

Figure 7.6 Winners of settlement farm competition.

Source: Hudson's Bay Company Archives

Table 7.1 Attrition rates of HBCOSL farm settlers

Year	Settlers	Returned to England	Left but still in Canada
1929	95	3	8
1930	99	3	1
1931	97	1	1
1932			
1933			
1934	83	4	6 (1 died)
1935	80	1	2
1936	76	2	3
1937	73	2	1
1938	68		5?
1939	64		4
1940	61		3
1941	57		4
1942	49		8 }
1943	38		11 }
1944	31 (3)		7 }
1945	30 (5)		12 }
1946	28 (7)		2 } (quit claims)
1947	23 (9)		5 } (1 foreclosure)
1948	22 (11)		1 }
1949	20 (11)		2 }
1950			
1951	18 (11)		2 }

( ) = paid for land and out of scheme

Source: HBCA

obviously no hope of eventual success (Table 7.1). Other departing settlers did so by quitclaim, a less complicated procedure that allowed the land to be resold more readily. In 1944, the first of 11 farmers made the final payment on their farm and became independent of the scheme. But by 1948, 78 farms had reverted to the HBC, most of which had been resold to other farmers. Just three years later, in 1951, the last year for which reports are available, only seven families remained in the scheme. Along with the 11 successful farmers, this amounted to a mere 18 properties, which meant that fewer than 20% of the hundred original farms were still in the hands of the HBC-sponsored settlers (Stuart 1949).

Although the HBC was firmly committed to its policy of recruiting only British migrants for the Vermilion farm settlement scheme, from where in Britain they were drawn appears to have been immaterial. Surviving company records occasionally identify the settlers' place of origin, but nowhere is this done systematically. Because of this, the passenger lists from ships carrying the settlers to Canada were used to identify places of origin. In all, 116 families were sent out under the Vermilion scheme before the Canadian government stopped its support of assisted immigration and the HBCOSL decided to take no more settlers (Chester 1931; Schultz 1990). The additional

Table 7.2 Status of farm settlers as of January 1938

Status	English	Other British	Percentage English	Unknown Origin
Not present	15	6	71%	1
Will not succeed	5	3	62%	5
Success is doubtful	9	5	64%	3
Totals	29	14	64%	11
May succeed	7	7	50%	3
Should succeed	8	5	61%	6
Will succeed	4	4	50%	2
Totals	19	16	54%	11
Overall Totals	48	30	61%	22

Source: HBCOSL Annual Report, 1938

families were brought as replacements for those who never reached their farms or who left almost immediately on arrival. Of the 116 families, 95 could be identified from ships' registers. Of these, 63% were English and 35% were Welsh, Scottish or Irish. It is noteworthy that of the 16 who left soon enough to be replaced, 13 were on the ships' registers, and 11, or 84%, of those were English (Stuart 1949). This higher failure rate for the English settlers was not just temporary. Table 7.2 shows that, as of January 1938, of the remaining 100 settlers whose origin are known, the English make up 71% of those who had left by that time. Of those rated as likely or certain to fail, 64% were English, whereas only 54% of those given a chance of success were English. By 1943, only four of the 15 English families who were still on their farms were given any hope of survival. Assuming those four did survive until the scheme was wound up in 1950, just over 6% of the English settlers, as against some 15% of the total number, saw things through to the end of the scheme. However, although HBCOSL officials referred to a number of the English settlers in less than glowing terms, they also did so in describing noncooperative settlers from other parts of Britain. For example, in the company's 1936 year-end report only five of the 13 settlers labelled as "misfits due to laziness or incompetence and for whom there seems to be no hope of successfully managing their own affairs as farmers" were identifiable as English (HBC 1936). The statistics show that the odds for success were slim wherever the settlers were from.

Needless to say, the HBC did little to prove that it could settle migrants on farms in Western Canada at a lower cost than the government. Indeed, the HBCOSL scheme fared much the same as the Canadian government's 3,000 Families Scheme, from which the great majority left their farms, with two-thirds of them leaving the region altogether and 11% of them returning to Britain (Smith 1981). In both schemes, the costs were huge, both in public and private outlay, as well as to the settlers themselves. At the end of 1942, the HBCOSL's agreement with the British government was terminated, with the government writing off losses of $100,000 and the HBC writing off $193,000 – perhaps an overstatement, as much of it was for land that had largely lost its value anyway (Rogers n.d.).

In its own evaluation of the settlement scheme, the company at times tried to claim it was a success in the sense that it brought permanent settlers to Canada, even though most of the settlers obviously were not going to succeed on their farms. One report asserted: "Back of every failure has been the settlers' warped view-point and wrong attitude – too prone to blame others and too lacking in initiative to dig in for himself" (Stuart 1949). This seems a rather harsh criticism in view of the natural disasters that affected the settlement, but other settlers in the district did manage to meet their repayments at a time when members of the HBCOSL scheme seemed determined to depend on outside support – in some cases blaming the company, alleging it was not forwarding the money the British government must surely be making available for them. There was certainly no wholesale abandonment of the district. Of the 79 farms that had reverted to the company by 1948, 68 had already been resold and only 11 remained in company hands (Rogers 1948). The record does not make it clear, but it seems likely that the real reason the company continued to try to make the settlement work was the recognition that resale of the land would not bring in as much as the debt still outstanding on the delinquent farms.

Conclusion

It is difficult to draw definitive conclusions from the small sample taken from this one limited scheme. Farm failures were widespread during the period – not just in Canada but as far away as Western Australia (Brayshay and Selwood 2002). Collapsing economies, adverse prices, ill-judged selection of migrants, lack of farm experience, inadequate farm size, poor knowledge of local conditions, insufficient capital, onerous debt loads and a host of personal problems stood in the way of success. However, some families did make a go of it, one suspects largely through access to personal capital or off-farm earnings. Some families certainly failed because they "lacked the will" and did not make sufficient effort, but given the circumstances, this would be an unfair judgement of most of the settlers. As for the English participants, their experiences did not appear to differ greatly from the rest of the migrant group, although it could be argued that their relatively high rate of early abandonment of the scheme was an indication of a lower level of commitment to the exercise. It does appear that some used the scheme merely as a vehicle to emigrate, but others, whatever their original intentions, were probably perceptive enough to realise that life on a prairie farm was not for them. They were certainly not alone in their thoughts. Conversely, some stuck it out to the end, outlasting the great majority of their kinsmen from across Britain. Under the extreme circumstances that prevailed during the dirty thirties, these outcomes are perhaps understandable. Nevertheless, the

HBCOSL's settlement scheme did nothing to dispel the widespread feelings in Canada that the British, and more particularly the English, were not the right choice of migrant to establish farms on the prairies. Unfortunately, the British settlers brought out under the HBCOSL scheme probably reinforced those negative sentiments.

References

Brayshay, M and Selwood, J 2002, "Dreams, propaganda and harsh realities: landscapes of group settlement in the forest districts of Western Australia in the 1920s", *Landscape Research* Vol. 27, pp. 81–101.

Brooks, JC 1929, "JC Brooks, secretary, governor and Committee Hudson's Bay Company [HBC] to Canadian Committee, 23 January", Public Archives of Manitoba [PAM], Hudson's Bay Company Archives [HBCA], Record Group [RG]2/7/216.

Brooks, JC 1930, "JC Brooks, Secretary HBC, to P.A. Chester, Chief Accountant, 7 January", PAM, HBCA, Hudson's Bay Company Overseas Settlement Limited [HBCOSL], RG2/7/216.

Canada Land Inventory (CLI) 1970, "Soil capability for agriculture; Vermilion 73 E" (1970); Wainright 73D (1970).

Chester, PA 1929, "PA Chester, chief accountant, Winnipeg, HBCOSL to Dominion Government Committee, 30 October", PAM, HBCA, RG2/7/216.

Chester, PA 1931, "PA Chester, general manager, to Governor and Committee, HBC, 5 May", PAM, HBCA, RG2/7/216.

Dennis, JS 1928, "JS Dennis, Commissioner, CPR to HF Harman, 18 September," PAM, HBCA, RG1/76/1.

Galbraith, JS 1951, "Land policies of the Hudson's Bay Company, 1870–1913", *Canadian Historical Review*, Vol. 32, pp. 1–21.

Garnet, WJ 1937, "WJ Garnett, report, 2 April", PAM, HBCA, RG2/7/218.

Harman, HF 1928, "HF Harman, land commissioner, HBC, to GL Bellingham, land agent, HBC, 8 June", PAM, HBCA, RG1/76/1.

Harman, HF 1928a, "HF Harman to Bellingham, 14 June," PAM, HBCA, RG1/76/1.

Harman, HF 1928b, "HF Harman to EH Gamble, Manager, HBCOSL, 16 June", PAM, RG1/76/1.

HBC 1928, "Telegram", Land Department to HBC, London, 29 May, PAM, HBCA, RG1/76/1.

HBC 1929, "Telegram", HBC London to Canadian Committee, HBC, 25 March, PAM, HBCA, RG2/7/216.

HBC 1930, "Annual report and correspondence of 100 Family Scheme", *Annual Report for Year Ending 31 January 1930*, PAM, HBC, RG 2/7/216.

HBC 1930a, *The Beaver*, December, p. 131.

HBC 1931, *Annual Report for Year Ending 31 January 1931*, PAM, HBC, RG2/7/216.

HBC 1932, "Report of Committee on 100 Family Scheme from 1 December 1930 – 31 March", PAM, HBCA, RG2/7/217.

HBC 1936, "British family farm settlement Vermilion area, general report for the year 1936", PAM, HBCA, RG2/7/218.

HBC 1938, *Annual Report for Year Ending 31 January 1938*, PAM, HBCA, RG1/76/3.

HBC 1951, "Annual reports and correspondence of 100 Family Scheme", from 31 January 1932 to 11 December 1951, PAM, HBCA, RG2/7/216-217.

HBCOSL 1926, "Articles of association", 17 September, PAM, HBCA, RG1/76/1.

HBCOSL 1928, "Agreement", 24 August, PAM, HBCA, RG2/7/216.

HBCOSL 1930, "Minutes of the Canadian Committee of HBCOSL", 29 May, PAM, HBCA, RG2/7/216.

Howard, AJP n.d., "AJP Howard to JS Dennis, commissioner, CPR", PAM, HBCA, A92/238/10.

Howard, AJP 1929, "Private", AJP Howard, Director, HBC to EH Gamble, Canadian Manager, HBCOSL, 14 January, PAM, HBCA, A92/238/10.

Howard, AJP 1929a, "Notes concerning future HBCOS policy", 26 March, PAM, HBCA, A92/238/10.

Howard, AJP 1929b, "AJP Howard to CV sale, Governor, HBC, 27 March", PAM, HBCA, A92/238/10.

Lupul, MR 1988, (Ed.), *Continuity and Change: The Cultural Life of Alberta's First Ukrainians*, Canadian Institute of Ukrainian Studies, Edmonton.

Nicholls, TH 1928, "Report No. 12", TH Nicholls, Assistant Manager, HBCOSL, 12 May, PAM, HBCA, RG26/4/11.

Peirson, FR 1932, "Canadian committee, HBC to Governor, 13 September", HBC, PAM, HBCA, RG2/7/217.

Rogers, CN n.d., "Summary of HBC Vermilion farm settlement scheme", PAM, HBCA, RG1/76/3.

Rogers, CN 1948, "Summary Vermilion farm settlement", 31 December, PAM, HBCA, RG2/7/217.

Sale, CV and Beatty, EW 1926, "CV Sale, Governor, HBC, and EW Beatty, President, CPR, to R Focke, Minister of Immigration and Colonization, Ottawa, 2 November", PAM, HBCA, RG1/76/4.

Schultz, JA 1990, "'Leaven for the lump': Canada and Empire settlement, 1918–1939", in S. Constantine (Ed.), *Emigrants and Empire: British Settlement in the Dominions between the Wars*, 1990, Manchester University Press, Manchester.

Smith, D 1981, "Instilling British values in the prairie provinces", *Prairie Forum*, Vol. No. 2, pp. 134–9.

Stuart, JM 1949, "British Farm Settlement Report", 15 November, PAM, HBCA, RG1/76/3.

Whishaw J 1929, "J Whishaw, Secretary, HBCOSL, to Secretary, HBC, 24 June", PAM, HBCA, A92/238/10.

8   Fordlandia and Belterra yesterday and today: lessons from failed projects of the Amazon

Ana Maria de Souza Mello Bicalho
Scott William Hoefle

Repeating the same mistakes

The Fordlandia company town is considered to be a textbook case for failed mega-maniac projects in the Amazon, much like the later Daniel Ludwig cellulose and rice production project in Jari, also located in Pará state, a cemetery for overly optimistic ideas about dominating the rainforest with advanced technical know-how and business savvy. However, not only foreign investors have faced difficulty in developing the Amazon. Brazilian planners also devised a number of settlement programmes that failed during the twentieth century for many of the same reasons. Today, success in developing farming systems that are environmentally sound, economically viable and socially inclusive is still as elusive as ever. The comparison of these experiences provides important lessons for understanding the serial failure of land settlement schemes in the Amazon.

The spectacular failure of the rubber plantation and the American cultural enclave of Fordlandia is a familiar story. The emphasis is usually on the impossibility of the task, and the faded epoch photographs presented only strengthen the image of abandonment and of an enclave of civilization reclaimed by the jungle. Today the site is visited by an occasional tourist who usually comes away haunted by the scenes of rusting structures and machinery, dust-covered vehicles and other period objects. The attempt to correct the experiment by creating another plantation in Belterra further downstream on the Tapajós River is treated in studies on the Ford experience but is less known by non-Amazon specialists and the general public.

The historical studies produced by Dean (1987) and Grandin (2009) correct this image. The two studies dovetail neatly to provide a well-rounded portrait of the Ford plantations. Dean uses ecological and political economical perspectives to depict rubber production in Brazil contextualised in global markets in which Fordlandia and Belterra are presented among many examples of the environmental and commercial failure of Brazilian rubber production during the twentieth century. Grandin takes a more cultural approach focusing almost exclusively on the Ford Motor Company and how the plantation experience expressed the technologically progressivist and morally conservative worldview of Henry Ford himself and so created social problems that also contributed to the economic failure of the plantations. In both studies, the immediate context of the two towns within the Santarém-Itaituba area is not treated well. Therefore, a middle course is taken here, situated between Dean's macro study and Grandin's micro study. This level of analysis can be called a "sub-regional", "areal" or "zone of influence" approach, which is used here to explain not only the different fates of Fordlandia and Belterra but also their parallels with later land use schemes in this part of the Amazon.

Furthermore, the Ford plantations gave rise to towns that long suffered neglect but were not engulfed by the unforgiving jungle. As the frontier has been consolidated in the Santarém-Itaituba area, the two have become small urban centres that are still served by much of the public infrastructure left behind by Ford in the 1940s. We will show that the plantations were indeed environmental, economic and maybe even social failures, as was the federal government planned colonization of the 1970s and 1980s. However, western Pará has changed over the last 20 years. Indeed, if he were still alive, Henry Ford might even be pleased to see his failures as tourist attractions; but the real driving forces for the rehabilitation of the two towns are those fuelling regional development. As this implies, the emphasis in this study is on the past in the present, what the two towns became rather than what they could have been.

Contextual methods

The Amazon is a vast region that is more than twice the size of Western Europe and possesses as much environmental and social diversity. This complexity is hard to capture using conventional regional analysis that relies too much on aggregate secondary data and deduction based on the personal biases of specific researchers. To counter this, numerous case studies have been made at the local level in different parts of the region, but all too often researchers still try to generalise from their particular experiences to the regional level. The aim of the present study is to place what seems to be the exceptional case of the Ford plantations in their immediate context: the Santarém-Itaituba area (Figure 8.1).

Drawing from historical accounts, scientific works, secondary data and first-hand research, we assemble trends from the bottom up in order to show parallels among the various settlement schemes of the twentieth century. First, the Ford plantations are placed within the historical context of the rubber boom and bust in the Amazon, which occurred shortly before the first plantation was installed. Then, using Dean (1987) and Grandin (2009) as well as research reports such as Cruls (1939), Camargo and Guerra (1959) and Castro (1973), which characterise Fordlandia and Belterra at different points in time, we briefly describe the rise and fall of the two plantations

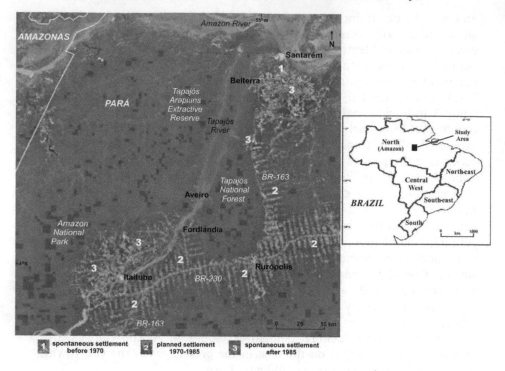

Figure 8.1 The Santarém-Itaituba study area.

Source: Adapted from Google Earth 2017

under the Ford administration and their ignoble fate afterwards. Finally, first-hand information from field research undertaken in the Santarém-Itaituba area in 2008, 2010 and 2013 is presented to explain the failure of the federal colonization projects of the twentieth century in western Pará state (for many of the same environmental, logistic, economic and social reasons) as well as the rural transformations of the last 20 years (which have breathed new life into Belterra but less so into Fordlandia).

Conclusions concerning land settlement in the Santarém-Itaituba area were constructed from the local level up by interviewing different rural actors who live in 39 communities located along the roads and rivers of the study area. In the field research, 126 riverine peasant, small settler, rancher and commodity farmer families were interviewed concerning their use of land and resources, farming systems, labour regimes, market articulation, sources of monetary income, family structure, out-migration, access to public services and political mobilization. During the course of the fieldwork, some elderly riverine farmers were encountered who had worked on the Ford plantations when they were young, and they are cited to confirm the views presented by historians. The researchers also talked with representatives of the Federal Environmental Protection Agency IBAMA (Instituto

Brasileiro do Meio Ambiente e dos Recursos Naturais Renováveis), the Federal Parks Service ICMBio (Instituto Chico Mendes de Conservação da Biodiversidade) and the Federal Land Office INCRA (Instituto Nacional de Colonização e Reforma Agrária) who work in the area as well as local leaders concerning conservation, land settlement policy, community development and partnership networks.

The boom and bust of rubber in the Amazon

The boom and bust of rubber tapping in the Amazon is well documented. Dean (1987) provides the best account of the economic and ecological history of rubber as a global commodity. First, based on a near monopoly of latex collected in the wild, Brazil took advantage of rising demand for rubber used in industrial applications at the turn of the nineteenth and twentieth centuries. Then Brazil fell victim to its limited capacity to supply global markets due to problems with fungi and insects that attacked rubber trees when they were planted on a larger scale. At the height of Brazilian dominance of the market, only 35,000 to 42,000 tonnes were produced per year, an amount which was exceeded by global demand as worldwide industrialization accelerated. Rubber tree seeds were taken out of Brazil, saplings were produced in botanical gardens and then transplanted to Southeast Asia where they thrived in plantation conditions far removed from their natural enemies in the Amazon River basin (Dean 1987, pp. 20–2).

In the study area, the city of Santarém, located at the confluence of the Tapajós and Amazon Rivers, played a minor role in the latex extraction boom because there were fewer rubber trees along the Tapajós River. This river does not have suspended sediments that form alluvial floodplains like those present along the Amazon River and its southern upper tributaries. The Tapajós River has crystalline water because its headwaters are located on the Brazilian Central Plateau, which has poor savannah soils. This limitation doomed the Ford plantations but not small-scale peasant production based on planting rubber trees in old fallows. Even with limited productivity, the commercialization of latex from the Tapajós River fuelled the growth of a medium-sized city, and the rubber buyers dominated local politics in Santarém during the late nineteenth and early twentieth centuries. Itaituba was founded by a member of this elite in 1870 to act as an advanced latex collection point located about 250 kilometres up the Tapajós River (Amorim 2000).

Santarém may not have had an important role in rubber tapping, but it did in the events that doomed this activity in the Amazon as well as in the failed attempt to mimic Southeast Asian plantation rubber production. For Brazilians, Henry Wickham is considered to be the villain who smuggled the rubber seeds out of the country, which permitted the setting up of the plantations in Asia that eventually monopolised global markets. Wickham was an Englishman who, in the 1870s, tried to make a living from producing sugarcane, tobacco and rubber on a farm located near Santarém. His cropping activities were not successful, but he gained experience with rubber production, and then took the seeds out of Brazil through the port of Belém with the full knowledge of the local authorities. These were taken to Kew Gardens and later from there to Malaya, where British capitalists set up rubber plantations as did Dutch planters with saplings transferred from the Buitenzorg botanical gardens to Sumatra and Java (Dean 1987, pp. 14–20).

Because Brazilian rubber trees suffered from the depredations of a number of insects and fungi, especially South American leaf blight (*Dothidella ulei*), they were dispersed to prevent cross infection. Furthermore, a slow and labour-consuming smoking process was used to cure the latex. Exports therefore fell off over most of the twentieth century. Not even specialised research over several decades was able to overcome these natural limitations. This is still a recurrent problem when monoculture is attempted in the Amazon (Dean 1987, pp. 58–60).

The failed Ford rubber plantations

It was in the context of the decline of Brazilian latex production that, in the 1920s, the Ford Motor Company decided to set up a rubber plantation in Fordlandia with the objective of breaking the British and Dutch monopoly on rubber. This was at a time, before contemporary flexible accumulation and outsourcing, when industrial firms pursued vertical integration whereby they strived to control all productive processes – from procuring the raw materials (producing rubber) to making parts (tires, hoses, etc.) and finally to manufacturing the final product on the factory floor (Ford automobiles and trucks). In addition to this economic rationale, Grandin shows that there were cultural and personal dimensions involving the plantations as experiments in social engineering and as a component of Henry Ford's rivalry with Theodore Roosevelt, which bordered on hubris.

The Ford plantations, like rubber tapping, were focussed on the rivers in historically settled Amazonia as opposed to the late twentieth century land settlement projects that converted rainforest into farmland along roads located far away from the rivers. The Ford plantations were fraught with environmental, economic and social problems from their inception. Instead of searching for a site in the upper Amazon, where natural conditions were better for rubber production, Ford was swindled by speculators and consultants into buying a location in the lower Amazon along the Tapajós River (Figure 8.2).

Ford originally paid US$125,000 for approximately one million hectares of land along an 80-kilometre stretch of the east bank of the Tapajós River. Even if this land could have been obtained for free, the price was not exorbitant given the size of the land concession. The problem was that the land was

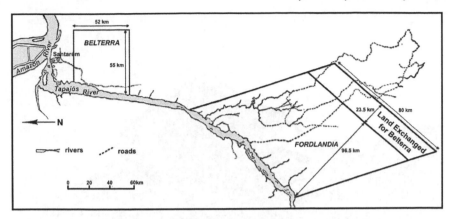

Figure 8.2 The Fordlandia and Belterra land concessions along the Tapajós River.

Source: Cruls 1939, p. 4

located in a part of the Amazon where poor, leached, sandy soils predomi-nate and where rainfall is highly seasonal, both of which slowed the growth of rubber trees. The terrain is also hilly, which made the use of equipment difficult and demanded more work from tappers. Finally, steamships took four days from Belém to reach Fordlandia and could not be used during the part of the year when the Tapajós was low (Cruls 1939, p. 4; Dean 1987, pp. 71–2).

Once the wrong site had been chosen in late 1928, everything was brought in from the United States to erect the plantation and company town. Comfortable houses were built for the American staff, first barracks for unmarried workers, then houses for married workers, warehouses, schools, a hospital, recreational facilities, an electrical generating plant and a water supply system, that is, urban infrastructure beyond anything available in the Amazon at the time except maybe in Belém and Manaus (Cruls 1939, pp. 11–4; Dean 1987, p. 73). The city was planned along an American grid scheme with individual houses located in the centre of lots that were distrib-uted in a block system rather than using the Brazilian model of compact urban settlement. The city exists to this day, and the electricity and water supply systems are still working while the hospital and warehouses fell into disuse and have become tourist attractions (Figure 8.3).

As Dean points out, the rubber plantation was much less of a success. The planting of the rubber trees went slowly. Seeds regularly failed; transplanted samplings died and had to be replanted a number of times. South American leaf blight is endemic in the Amazon and flourishes in plantation farming systems in which trees are planted close together because this facilitates the spread of the blight from tree canopy to canopy. In the wild, *Hevea brasiliensis* rubber trees grow alone and are scattered over large tracts in order to avoid

Figure 8.3 The company town legacy in Fordlandia.
Source: Field research 2013

contact with the blight. During the first years of the plantation, no one on
the staff had scientific training in tropical agriculture or even in rubber production. The blight appeared even before the trees reached maturity for production, which normally takes seven to eight years, and attained epidemic
proportions at Fordlandia after 1935. One reaction was to set up another
plantation further down river in Belterra in 1934 where soil conditions on
the Santarém plateau were better and the land was level, higher and drier.
Belterra also had the advantage of having a year-round, deep-water port
located near the confluence of the Tapajós River and the Amazon River.
All of the same excellent urban infrastructure of Fordlandia was replicated in
Belterra (Dean 1987, pp. 76–7).

Another reaction was to try to use more resistant and productive *Hevea*
clones developed in Southeast Asia where local fungi did not attack the
introduced rubber trees. During the 1930s, Belterra became a centre of
research and experimentation in trying to adapt different kinds of clones and
grafting methods to Amazonian conditions, and this centre regularly traded
material with other research centres in the Americas and in Asia but to no
avail. The groves in Belterra also eventually became infested with blight and
production plummeted. In late 1945, with the end of World War II, Ford
gave up. After having invested more than US$20 million, the plantations

were sold to the Brazilian federal government for the value of US$250,000, which, according to Brazilian labour laws, was what was owed to workers in severance payments. Some workers went to live in nearby communities while others moved to other parts of the Amazon or returned to the Northeast (Dean 1987, p. 105).

The plantations were taken over by the Ministry of Agriculture and maintained for some years as research stations with a limited staff. Researchers repeatedly claimed to be on the verge of developing resilient rubber trees, and different reports over time, such as Camargo and Guerra (1959, p. 15) and Soares (1973, p. 8), stated that this had in fact occurred or was about to take place. They and the staff shared the same modern scientific worldview and were confident that this would indeed happen. However, the South American leaf blight was never eradicated. In an age before gene splicing, it was concluded that separate sets of genes controlled productivity and resistance to blight and that each worked at cross purposes so that only low-yield rubber trees showed any resistance (Dean 1987, pp. 96, 117, 143). Even today, the blight remains a barrier to plantation rubber production in Central and South America. Asia still supplies 90% of all latex produced in the world but is constantly on its guard to avoid the introduction of leaf blight into plantations located there (FAO 2011).

Of the three research stations along the lower Tapajós, two were closed, and the Belterra station eventually became the EMBRAPA forestry research unit. The latter still works with experimentation on timber production. Most of the land of the two plantations was incorporated into the Tapajós National Forest, and the rest was turned into land settlement schemes or was occupied by squatters (Dean 1987, p. 159).

The environmental impact of the Ford plantations was enormous and could have been much worse if some of the more hare-brained suggestions for combating the blight had been followed. To build the towns and to plant thousands of hectares of rubber trees, huge stretches of forest were cleared and burned. The cover of the book *Fordlandia* shows the forest coming right up to the backs of the houses aligned along an American suburban street. While this might have been good for selling books in the United States, the reality was quite different. Extensive deforestation occurred and a domesticated landscape created in its place. Cruls' article and Grandin's book are richly illustrated with period photographs and, apart from the planted rubber trees, hardly any other trees are to be seen (Figure 8.4).

Fungicide was used to protect the slices made in the lower trunks of rubber trees for extracting latex, but the trees were too high and canopies too close together to apply pesticides in the upper part of affected trees. Dubious proposals suggested using preventive aerial spraying of fungicide as well as herbicide to defoliate affected trees (Dean 1987, pp. 135–36). Today these practices are unacceptable in the Amazon, and, in this regard, it was probably better that plantation-scale rubber production failed.

Figure 8.4 Deforested domesticated landscape of Fordlandia.

Source: Ford archives 1933

Dean did not consider labour relations to be a problem for the success of the plantations even if American supervisors could be insensitive. Ford paid wages that were higher than the dismally low going rate for rural workers in the Amazon. The main difficulty encountered by Ford was the scarcity of labour. At the height of the rubber boom, there were 150,000 workers in this sector alone, many of them concentrated far away from Fordlandia in Acre state. With the decline of the latex gathering economy, many of the North-easterners left the Amazon. By 1930, there were only 250,000 male inhabitants left in the whole region. Recruiters searched for workers but, unlike the rubber buyers, Ford did not pay the expenses to bring the workers to the plantations. These candidates might not even be selected, in which case they received no help for the trip back to their place of origin. This treatment might make sense in the automotive sector in the United States, where workers lived in nearby urban areas, but not in the Amazon (Cruls 1939, p. 15; Dean 1987, pp. 81–2). One elderly man interviewed in a community located across the river from Fordlandia, who had worked there for ten years in his youth, had no complaints to make about the working conditions and stated that he liked the wages and was sorry to see the plantation close.

Grandin (2009) focused his account on the cultural and social side of the Ford experiment and specifically on Henry Ford's worldview. He traced the idea of establishing a plantation town in the middle of the Amazon to Ford's rivalry with Theodore Roosevelt. The latter had a river named after himself, which he had "discovered" when exploring in the western Amazon, so Ford wanted a city with his name (pp. 49–53). After a bad social beginning, which resulted in worker riots in 1930, Ford embarked on an experiment in replicating the public and private space of a small Midwestern town in the Amazon, complete with picket fences and red fire hydrants.

Grandin provides a fascinating and well-illustrated account of this planned town with neat little white houses complete with yards and gardens. He also points out that the planners were often out of touch with reality. For sanitary reasons, the houses had cement floors and tin roofs lined with asbestos. The roofs were designed by Ford engineers to reflect sunlight but instead kept heat in and caused the houses to be sweltering. The local people preferred thatch roofs and dirt floors because that kind of house was cooler (2009, pp. 272–5). However, Grandin had never lived in such a house, as one of the authors has. All kinds of insects rain down on the inhabitants, come out of the wattle-and-daub walls at night and live in the dirt floor causing a number of health problems. A fixed town of thatched-roof houses of Fordland's size would also deplete an enormous area of the species of palm trees used for thatch, which would then not be available when the thatch needed to be replaced. This kind of housing is adapted to conditions of low population density, such as in the past when subsistence communities shifted sites periodically in order to access new resources.

A town that had Ford's name also had to promote moral improvement and progress. Married men were recruited. Brothels, bars and gambling houses were removed. Wholesome leisure activities were encouraged to keep Brazilian and American workers and their families occupied and out of mischief. Football (soccer) was the preferred form of after-hours recreation as it is in all of Amazonia to this day. Playgrounds were built for children. Tennis courts and a golf course were constructed to practise sports that encouraged individual participation. This kept people apart so as to avoid creating crowds that could turn violent. Members of the American staff were also told to go hunting and fishing. Radio reception was relayed live from the States to Fordlandia.Vaudeville shows were staged. An open-air dance hall was built where traditional American square dancing took place while sensuous contemporary dances of the period, such as the charleston, foxtrot, rag, black bottom and tango were frowned upon. Films were shown, and the workers enjoyed American B action pictures and Brazilian slapstick and musicals (Grandin 2009, pp. 278–88).

Since the workers had families, babies were born and children had to be raised. This led to health care provision, courses on nutrition and personal hygiene, day-care facilities and schools. Workers died and had to be buried. They left widows behind who were given a pension. These expenses ballooned at a time when the plantations did not turn a profit (Grandin 2009, pp. 291–3). In the end, the towns only seemed to be a success to visitors like Cruls (1939), who marvelled at the infrastructure, or to Dean (1987, p. 73), who focussed on the environmental limitations and issues of global economic competitiveness in rubber production. In other words, Fordlandia and Belterra were failures on all counts. The difficulties encountered with providing social services also plagued the large-scale settlement programmes of the second half of the twentieth century and were of the same magnitude as the environmental and economic issues.

Failed federal planned settlement

Three types of settlement policies were attempted in the Amazon during the twentieth century: (1) small-scale model farming colonies established by the federal government from 1930 to 1970, (2) large-scale federal planned settlement projects set up from 1970 to 1985 and (3) limited decentralised state and local initiatives undertaken after that date when most planning became reactive to the land conflicts and deforestation caused by the spontaneous settlement that was occurring in a neo-liberal political vacuum.

Prologue: agricultural colonies and first road settlements

From 1930 to 1970, a limited number of farming colonies were set up along the Amazon River, usually near important cities, as part of a federal policy meant to create enclaves of smallholders who would plant vegetables, fruit and other specialty crops as well as raise small livestock meant to feed nearby urban areas. These projects often involved Japanese immigrants who had experience in these farming activities. Colonies in the Amazon were established on high ground and not on floodplains where mestizo peasant farmers have historically produced traditional crops such as maize, manioc and bananas. Three such colonies were set up in the immediate hinterland of Manaus, another near Parantins further down the Amazon River and others near Belém. The inland Japanese settlements at Tomé Açu and Castanhal were the most famous of these colonies. No colonies were set up in the study area, perhaps because three federal research stations already existed there. These, like the agricultural colonies, were supposed to provide technical innovations that would be adopted by local farmers, but this rarely happened because these methods were often inappropriate to local environmental and social conditions.

In addition to this limited federal settlement policy, during the 1950s and 1960s, some settlement took place up to 60 kilometres south and east of Santarém along rudimentary roads built by the Pará state government. Merchants from the city set up cattle ranches along these roads, and poor peasants from the Northeast squatted on the land. The latter only eked out a living from subsistence agriculture because cropping activities were restricted by the poorly maintained roads (IBGE 1970). Farmers could not get their produce to market during the rainy season when most crops were harvested while ranchers did not have this problem. Cattle could be herded along a muddy track that a truck could not pass through, or ranchers could simply wait for the end of the rainy season and take their animals to market when they were prime for sale. This basic transport problem doomed many of federal small-settler projects implemented in subsequent decades. The newly constructed highways were not maintained properly, and therefore crops could not be marketed. Significant urban markets were few, located far away from the settler projects and attempts to introduce cash crops, which could be sold elsewhere in Brazil or exported abroad, also failed for many of the same reasons that Fordlandia failed.

The heyday of planned federal settlement projects

The top-down, large-scale planned settlement of the Amazon undertaken by military governments from 1970 to 1985 is a well-studied subject (see Ianni 1979; Foweraker 1981; Hecht and Cockburn 1990; Schmink and Wood 1992; Oliveira 1999). Promoting greater settlement off the main rivers was part of an overall policy of regional development meant to increase population in the Amazon, to integrate the region into the national economy and so avoid geopolitical threats to political sovereignty over the region. Major highways were built in a north-south orientation, connecting important cities of the Central-West region to the Amazon and the famous Transamazonian Highway (BR-230), which cut across these highways with the intention of tying them together and providing a conduit for out-migration of poor peasant farmers from the underdeveloped Northeast region. The whole process was considered as a means of colonising a region considered to be empty of people, as if the Amerindians and the historic mestizo peasantry, who had been living there for centuries, did not exist. This had grave consequences for these social groups (see Survival International 2000; Bicalho 2009, 2010).

Settlement projects were set up and directly administered by the Federal Land Office INCRA, mainly along roads located in the eastern and southern Amazon. In the study area, settlement occurred along the east side of the BR-163 south from Santarém to the new model colonization town of Rurópolis and from there along the BR-230 to the Tapajós River near Itaituba. Feeder roads fanned out from the highways, creating a fishbone pattern of land use. Small farmers were settled on 100-hectare lots, extending 200 metres along the roads and 500 metres back from the roads into the forest. Behind the small farmer projects, large farmer projects were supposed to have been set up but never eventuated.

In the beginning, a productivist mentality prevailed; farmers were expected to clear and put 50% of their land into agricultural production while the other 50% was to remain in forest. In addition to food crops, planting cacao and black pepper was encouraged in order to generate income. Cacao was a traditional cash crop in riverine areas of the Amazon and, since the late 1940s, Japanese immigrant farmers had successfully

146 Ana Maria de Souza Mello Bicalho and Scott William Hoefle

planted black pepper on a commercial scale in settlements located off the rivers in eastern Pará. These crops could be readily sold in national and international markets and were not basic food crops that had to wait for the development of urban markets in the Amazon. Later, the regional fruit tree cupuaçu was planted on a commercial scale with the same purpose.

However, as we saw previously with rubber trees, when planted on a larger scale, all three tree crops developed problems with fungi and bacteria, particularly in the study area. Interviewed riverine farmers cited a failed attempt by a large farmer to plant cocoa on a commercial scale on the opposite side of the river from Fordlandia in the 1940s, which showed that fungi presented a problem with planting this crop in land with low fertility. Similar problems plagued black pepper. Just before the large-scale settlement programme began in 1969, black pepper planted by Japanese farmers began to face enormous problems with *fusarium* bacteria, which eventually caused many of the farmers to abandon their land, leave Tomé Açu and go to live in Belém (Animazon 2017).

A few nature reserves were created in the Amazon, but these were fig leaves meant to reduce criticism of the large-scale forest destruction that was occurring. The Amazon National Park and the Tapajós National Forest were two of the most important conservation units created at the time. Both are located in the study area and were subject to invasion by small farmers and large ranchers, a problem that plagues these conservation units to this day.

Following a strategy of integrated development, farmers were organised into communities where they were provided with a house and where farm extension and school and health services were to be made available. As the provision of these services to all of the settlers in such a vast region was costly, only a few farmers living in showcase projects like Altamira and Rurópolis actually received them locally. Most farmers out on the road heads were often abandoned to their own fortune where they could remain isolated for months during the rainy season.

Moran (1981) provided a detailed study of Altamira, which was far more successful than Rurópolis, which is within our study area. When the projects were instigated in Altamira, it was already the most important city along the Xingú River. Better soils were available there, and cacao cropping was less vulnerable to fungi. Altamira is located on the eastern section of the Transamazonian Highway where regional integration eventually occurred. By 2010, the city had a population of 84,092 inhabitants (IBGE 2010); it has been reasonably well served by roads in recent decades and has an airport served by jet aircraft. Altamira also has the dubious destination of being the murder capital of Brazil. The city has a homicide rate of 125 per 100,000 inhabitants, which is four to five times the rate in the violent metropolitan areas of Southeast Brazil. This has much more to do with the construction of the Belo Monte dam, the presence of rival drug factions in the city and its role as a thoroughfare than with classic frontier disputes over land ownership as in the past (cf. Hoefle 2006; Simmons et al. 2007; Nogueira 2017).

Rurópolis was created from scratch and was supposed to take advantage of its location at the junction of the BR-163 and BR-230 highways. For decades, local sections of the two highways experienced enormous problems with road maintenance, and Rurópolis remained a small town with a depressed rural-based economy. By 2010, there were only 15,273 inhabitants living there (IBGE 2010), but this situation may change because the Cuiabá-Santarém highway is being paved – in which case, it may finally become the thoroughfare that it was once planned to be. This change will be accompanied by new economic opportunities but also by urban problems like those now present in Altamira.

Alongside smallholders, ranchers occupied land along highways located between the feeder routes, bought up land from unsuccessful smallholders and occupied land where small settler projects were supposed to have been set up but never were. The ranchers were usually merchants from nearby towns, migrants from southern Brazil who made a career of moving along the frontier and preparing ranches for sale, and speculators from the larger cities of the Amazon as well as from other regions of Brazil. The latter often received subsidies from the federal government with the objective of introducing agribusiness into the region, but instead of this, the recipients diverted the subsidised loans into speculative investments in the financial market. The whole process was accompanied by usurpation of land and violent conflict was endemic (Schmink and Wood 1992; Oliveira 1999; Hoefle 2006; Simmons et al. 2007; Hecht and Cockburn 2011; Aldrich et al. 2012).

Aftermath: spontaneous settlement after 1985

With the end of military rule in 1985, planning became more decentralised, if it was undertaken at all, and the result was a wave of unorganised spontaneous settlement that fuelled land conflict. In a chaotic context of hyperinflation, recession and political instability, land conflict and deforestation escalated in the Amazon. Strapped for funding, INCRA gave up on integrated development projects and limited its action to granting land titles in the hope of defusing conflict over farm ownership. So-called projects were set up after the consummated fact of spontaneous settlement. Settlers would initiate the process by cutting a path from a highway into the forest to access new land and eventually opening a rudimentary road that INCRA could later improve. At one point, a small grant of money was given that was supposed to help settlers during their first two years. However, when the settlers received the stipend, they would immediately buy a chainsaw, cut down and sell all of the valuable trees and use the money for farming and living expenses.

The environmental consequences of all of this was a surge in deforestation that the newly created environmental protection agency IBAMA tried to control. To pre-empt deforestation, a large number of conservation units were set up from the 1990s onwards. A 1998 law reduced the amount of land that could be cultivated to 20% of a farm, which only made frontier farming more difficult.

This is clearly seen in the study area. The typical fishbone pattern of planned land settlement is encountered along the BR-163 and BR-230 between Santarém and Itaituba, while to the west of this city, ranchers usurped land in a 40-kilometre radius that had originally been earmarked for small-settler projects that were never implemented. Beyond that point, small settlers arrived after 1990 and installed themselves on land situated behind the ranches. Some of them cleared land located within the Amazon National Park as did others on the eastern edge of the Tapajós National Forest. The area closer to Santarém and Belterra was already more deforested and experienced settlement prior to the federal projects.

A crucial difference between the study area and areas located further east and south in the Amazon was the Itaituba gold rush, which slowed the local rate of settlement until 1990. The immediate impact of the gold rush on population movements, besides attracting outsiders directly to the camps, was to empty the planned colonization projects being established along the new highways at that time. The gold rush collapsed in 1990, and this gave a boost to agricultural settlement. Some prospectors went to the gold fields of Roraima, but others stayed on in western Pará where they settled on farms located along feeder routes opened on the advancing frontier (Mathis 1998; Hoefle 2013).

Today, there are three kinds of small settlers in the Itaituba-Santarém area: (1) classic frontier peasants who live in failed projects and areas of spontaneous settlement located far from consumer centres along rudimentary feeder routes connected to poorly maintained highways, (2) settlers in consolidated projects located on feeder routes and on the BR-230 close to the city of Itaituba where roads are reasonably maintained and a modest consumer market exists nearby, and (3) relatively prosperous settlers who are situated along the paved BR-163 near Santarém and have access to a larger consumer market.

The first group is composed of ex-prospectors who became frontier peasants along the advancing frontier of western Pará and who suffer the typical limitations to commercial farming experienced along most roads in the Amazon. Crops are harvested during the rainy season when roads become mired in mud and traffic becomes impassable. Therefore, farmers do not produce more lucrative, perishable crops and instead plant modest areas of beans, maize, manioc and fruit for self-provisioning and rice for local markets (Table 8.1).

Pigs and poultry are raised mainly for local consumption, and some cattle are sold in the limited urban markets of western Pará, where cities are few and far between. However, the main cash activity, rice production, is being curtailed due to the appearance of a fungus that destroys the crop and again demonstrates the perennial problem that occurs when cropping is attempted on a commercial scale in the Amazon. Consequently, little

Table 8.1 Land use of small farmers situated along roads in the Itaituba-Santarém region

Type of Farmer	Cropping		Pasture for Cattle		Short or Long Fallows		Forest		Total	
	ha	%	ha	%	ha	%	ha	%	ha	%
Frontier peasants on road heads	4.6	5.4	7.2	8.5	3.3	3.9	69.5	82.2	84.6	100.0
Farmers near Itaituba	6.4	5.0	7.2	5.7	9.9	7.8	103.3	81.5	126.8	100.0
Fruit farmers near Santarém	3.2	27.8	2.0	17.4	5.3	46.1	1.0	8.7	11.5	100.0
Poultry farmers on paved roads	1.8	1.8	0	0	38.2	38.2	60.0	60.0	100.0	100.0

Source: Field research (2008)

monetary income is earned by these peasants, and the result is dependence on government transfer payments (Table 8.2).

Farmers on the road heads have very little land in fallow – less than the area they plant – and this represents a serious threat to sustainability over time. No fertilizer or any other means of controlling pests and crop disease is used so production is based solely on the application of human labour and the natural fertility of the soil. After an area of primary forest is opened up for cultivation, fertility is rapidly depleted. According to classic studies of tropical shifting agriculture, in order to renew fertility, land in fallow should

Table 8.2 Annual average income by source for workers and small farmers situated along roads in the Itaituba-Santarém region (US$2008*)

Type of Farmer and Worker	Cropping	Cattle	Poultry	Rural Wage	Non-rural Wage	Transfer Payment	Total
Crop day worker road heads	0	0	0	470	0	244	714
Salaried cowhand road heads	0	0	0	2,120	0	0	2,120
Small farmer road heads	408	378	93	8	385	833	2,105
Small farmer near Itaituba	3,103	475	397	0	297	1,250	5,522
Fruit farmer near Santarém	5,313	0	221	0	989	0	6,523
Poultry farmer paved road	0	0	16,630	0	0	0	16,630

*Annual Federal minimum wage 2008 = US$2,346.

Source: Field research (2008)

be at least five times if not eight times that of the land in crops (Boserup 1965; Ruthenberg 1980; Simmons 1989).

To overcome these problems, some frontier farmers try to raise cattle primed for sale after the rainy season, but a standard 100-hectare lot, of which, since 1998, only 20 hectares can be cleared for use, does not furnish enough pasture to support a herd large enough to generate a decent living. Farmers end up clearing more land for pasture than that permitted by law and thereby run afoul of IBAMA. To earn money, peasants can also resort to selling hardwood and animal skins, which obviously are not sustainable practices. Parcels of land can be sold, and the substandard size of their farms reflects this. On average, the frontier peasants interviewed had 84 hectares of land, and further sale of land parcels would only make their situation worse.

Lack of sustainability in frontier farming in turn causes high rates of spatial mobility as farmers move on when the natural fertility of the poor soils of the study area is exhausted. Like ranchers, many peasants have made a career of moving along the frontier. Fifty-seven percent of the peasants interviewed have moved five to eight times during their lifetime and another 16% from nine to fifteen times. Many of the moves took place in the past due to the high mobility of gold prospectors going from camp to camp, but exhaustion of the soil through lack of fallowing has been responsible for most spatial mobility since 1990.

However, this historical pattern of shifting along the advancing frontier was interrupted when the peasants reached the poorly demarcated limits of the Amazon National Park. Clearing land in the park provoked a dispute between the farmers and their labour unions on one side and the environmental protection agencies on the other. The park ended up losing some land in the east and gaining more land further west, but now the eastern limits of the park are more visible. The Tapajós National Forest also lost a small area to settlers along the BR-163 in the locality of São Jorge (see Figure 8.1).

This cat-and-mouse game could have gone on until the whole Amazon was deforested, but the political context has changed over time. In recent decades, the environmental protection agencies and their political allies have become more powerful and are better equipped with real-time satellite imaging that permits agents to quickly detect infractions. Nevertheless, there are a host of local, regional and national vested interests in logging, land speculation, building hydroelectric dams and political corruption that act to limit the efforts of IBAMA and ICMBio.

The second group of farmers, located in areas of consolidated settlement near Itaituba, encounter fewer problems than those experienced by the frontier peasants of the road heads. Farmers in the second group use reasonably well-maintained unpaved roads and have access to a nearby urban market with more than 70,000 consumers (IBGE 2010). With an average farm size of 126 hectares, they own over 50% more land than farmers on the road heads.

As a result, they have more land in cropping and pasture, which provides better prospects for selling produce, and they earn six times more income from rural activities than do the peasants. Like the peasants, all of them were once gold prospectors, though this prior experience in itself does not make them unsuccessful farmers. Similar to the "stickers" of the US frontier literature (cf. Hine and Faragher 2000), they constitute the minority of frontier farmers who are successful. They have lived on their current landholdings for an average of fifteen and a half years and, before this, 83% had moved fewer than four times in their lifetimes.

The third group of smallholders are located along paved highways near Santarém within what became a commodity-production landscape after 2000. These small farmers have year-round access to a sizeable urban market for fruit and chickens with more than 210,000 consumers (IBGE 2010). They have access to mains electricity, which is used to aggregate greater value by allowing them to sell fruit pulp and undertake henhouse poultry production. Their close proximity to Santarém also provides opportunities for employment in town or locally as county workers in schools and health facilities. Consequently, these farmers earn more agricultural and non-agricultural income than do the farmers located further away from Santarém. They have relatively prosperous livelihoods and do not depend on government transfer payments. As such, they have the life that farmers of the distant failed settlement projects could have had if market and logistical limitations had been overcome.

Ranchers complete this cast of characters of rural actors present in the post planned-settlement landscape. Some ranches still exist south of Santarém, but most are located beyond the commodity landscape further south in Belterra and Aveiro municipalities. The owners interviewed here originally came from the Northeast. They live in Santarém and rose through the ranks of petty urban merchants to success in larger commercial establishments, such as supermarkets. The ranch represents a secondary line of investment for them.

Ranching is land extensive and needs large areas of pasture to generate significant income. One ranch visited in 2008 was situated about 100 kilometres south of Santarém at the end of what was then the paved part of the BR-163 highway. The ranch had a total area of 2,800 hectares with 300 hectares of grazing land for 300 head of cattle. Steers were sold directly to a meat packing plant in Santarém for a net income of US$50,898 in 2007. The owner estimated that the ranch generated about 5% of all of his income. His ranch hand earns the minimum wage, which was US$2,791 a year in 2007.

A small ranch owner interviewed in 2013 had 100 hectares of land located north of Itaituba. This little ranch demonstrates how frontier stock raising is more economically viable as well as more environmentally threatening. All of the land had been cleared, with 88 hectares in pasture and 12 hectares in cupuaçu trees grown under the shade of rubber trees. The rancher has 120 head of cattle and sells between 20 and 30 animals a year. In 2013, he earned US$9,094 from stock raising and no income from cropping. Up to now, the shade of the rubber trees has prevented the appearance of fungi in the cupuaçu trees, but the ranch does not have the electricity necessary for converting cupuaçu into fruit pulp for sale. No latex is extracted from the rubber trees.

Much of the area in ranches around Itaituba is owned by a single individual who is also a powerful local politician. IBAMA and INCRA officials complain that, over the years, the rancher usurped and deforested more than 500,000 hectares of land originally destined for small-farmer projects and the Amazon National Park. In the past, this kind of social actor would be in direct conflict with the smallholders, but he did not oppose the frontier peasants squatting on the land at the back of his ranches. As a result of this development, he was able to claim that it was they who had occupied the land that should be in forest. In the changed political context of Brazil today, rather than being natural enemies, the rancher and the peasants have actually become political allies against a common adversary: the environmental protection agencies. This alliance was concretely expressed in the 2008 elections when the rancher was almost elected mayor of Itaituba.

Commodity versus conservation landscapes after 2000

Today, the general region that includes Fordlandia and Belterra has been transformed by the introduction of commodity production as well as by environmental policies for the Amazon that are meant to preserve natural areas against the expansion of commodity production, logging and ranching. The northern part of Belterra municipality and the town itself are now within the deforested landscape of commodity production, which is gradually expanding southwards into the area of small settlers and ranchers located outside the Tapajós National Forest on the eastern side of BR-163. Most of the southern half of the municipality is situated within the preserved landscape of the Tapajós National Forest where innovative community-based sustainable timber and non-timber forestry is undertaken. The two landscapes express productivist and post-productivist development strategies, respectively, one based on settlement along roads and the other involving historical peasant settlement along rivers.

Fordlandia is located just south of the national forest that, together with the Tapajós-Arapiuns Rubber Tapper Reserve on the opposite side of the river, occupies almost the whole area of Aveio municipality. However, Fordlandia has not participated in either of the two innovative rural landscapes to the north. The settlement projects nearby are now derelict. One interviewed family in a riverine community lives there in order to have access to commerce and social services, but they also have farmland in this derelict project. The husband spends his weekdays on the farm where he works his land alone. Their only remaining neighbour is a small rancher. All the rest of the settlers have left to live in Itaituba. The closed local office of the national centre for cacao production in Fordlandia reflects the collapse of cash cropping on the nearby projects. Only poor riverine farming along the Tapajós River remains.

Beginning in the late 1990s, commodity producers started to buy the ranches and small farms on the Santarém plateau, which presented better conditions for soy production than those normally encountered in the Amazon. The land is relatively flat and not crosscut by watercourses, and this permits the use of mechanised commodity production. Prior consolidated land ownership in the ranches also allowed commodity farmers to buy large parcels of land in order to attain the minimum scale necessary for the use of large machinery. Cargill built a port in Santarém that reduced transport costs and made it possible to obtain higher soy prices.

Before arriving in Santarém, almost all of the commodity farmers were originally from the South and had farmed in the Central-West region, Brazil's major zone of agribusiness. What Wilson and Burton (2015) call highly specialized super productivist farming is now being undertaken. Large planting and harvesting machinery, lime, chemical fertilizers and pesticides are used extensively. Grandin correctly criticises the environmental and health problems that accompanied the introduction of this kind of farming near Belterra, but he exaggerates the case for technological determinism destroying the Amazon. Commodity production leapfrogged directly to Santarém and is not expanding unfettered up the BR-163. Exceptional conditions exist for commodity production in the study area, which do not exist in much of the Amazon. When Grandin was writing his account of Fordlandia and Belterra, an alliance of local, regional and international environmental groups reacted to this expansion of commodity production and staged a successful protest that pressured the federal government to enforce environmental laws. Commodity production fell off and then returned to previous levels, oscillating between 27,000 to 30,000 hectares planted in the years since then (IBGE 2017). See Hoefle (2012) for greater detail concerning environmental, economic, logistical and political limitations to commodity production in the Amazon.

The amount of land in commodity production in the study is still modest when compared to the rest of Brazil. Farmers interviewed in the Santarém-Belterra area planted from 120 to 1,000 hectares of soy, rice and maize in 2008. The smallest farmer earned US$21,600 that year and the largest US$53,913. The smallest farmer worked his land mainly using family labour aided by machinery and occasionally contracted day workers at the rate of US$13 a day. The largest producer had a more complex division of labour: a foreman who earned US$6,435 a year, two tractor drivers who earned US$3,518 a year and three seasonal workers who earned US$180 a month for half of the year. In classic semi-proletarian fashion (cf. Kautsky 1988), workers are recruited from the families of small farmers who live nearby and then trained. Given this hiring practice and the presence of the fruit and poultry farmers mentioned earlier, it is an open question as to whether smallholders will be automatically eliminated from this landscape.

Against this productivist landscape, a preserved landscape exists in the Tapajós National Forest, where multifunctional non/post-productivist rural activities are undertaken (cf. Wilson and Burton 2015). The northern limits of the national forest are located about 50 kilometres south of Santarém and near the town of Belterra. The eastern side of the conservation unit runs along the BR-163, and the southern boundary is just north of Rurópolis. The western boundary of the national forest is the Tapajós River where peasants and Amerindians have lived for centuries. What was once the southern part of the Belterra Ford plantation and the northern part of the Fordlandia plantation are now within the national forest.

A logging management zone exists in the national forest where a cooperative representing the riverine communities undertakes sustainable forestry. A sector-rotation system was initiated in 2009 over an area of 32,222 hectares. The management plan permits the logging of one 1,000-hectare sector per year after which the sector is left undisturbed so that forest regrowth can take place over a 30-year period.

Today, this forestry system is operated by the local rural population and not by commercial firms because of a contentious 40-year struggle for rights in which the strengthening of collective goals produced the social capital necessary for the undertaking of community-based forestry. The social movements and political organization of the local population that arose from their struggle for land and resource rights involved a long process of community organization, culminating in the Tapajós National Forest-wide Cooperative being set up to undertake sustainable commercial logging. In 2012, an institutional conflict with the Brazilian Amerindian protection agency FUNAI (Fundação Nacional do Índio) resulted in the loss of half of their logging area. However, the cooperative was compensated with another area more than twice as large located further south in the national forest so that, beginning in 2014, the logged area was doubled to 2,000 hectares a year and the work force increased accordingly. With a view to creating more work and income, the cooperative has plans to build a sawmill in order to aggregate greater value than that earned from selling brute logs, and this will create jobs for another 50 workers.

As a result of this change, families with members employed by the cooperative will grow from 5% to 14% of the total number of families living in the national forest. Induced by environmental agencies and foreign governmental and non-governmental organizations (NGOs) who favour low-carbon emission, non-timber forestry over logging, the collection of Brazil nuts, natural oils and tree seeds, the extraction of latex, the making of wood handicrafts and furniture and the undertaking of ecotourism have been developed by associations and individuals in the community area of the national forest. These activities are based on local knowledge and skills but are relatively small-scale and do not generate much income. Due to their environmental and cultural appeal, these activities are highly encouraged by the NGOs, but they are not given priority by the cooperative, which has a better sense of what is commercially viable.

Table 8.3 Average income by source for workers and small farmers of the Tapajós River (US$2013*)

Rural Activities	Rural Worker	Farmer Outside National Forest	Farmer Inside National Forest	COOMFlona Cooperative Worker
Crops	0	663	678	3,406
Small animals	0	34	22	58
Cattle	0	103	0	235
Latex & natural oils	0	396	138	550
Fishing	0	1,658	69	0
Wages	3,980	0	89	8,593
Pension & family support	1,101	1,778	3,421	903
Environmental services	0	0	23	550
Total	5,081	4,632	4,440	14,295

*Annual Federal minimum wage 2013 = US$4,043.

Source: Field research (2013)

However, the remainder of the population in the national forest do not enjoy the high income that the lumberjacks receive, and their farming activities are hampered by environmental restrictions (Table 8.3). Farmers within the national forest are not allowed to plant more than one hectare of land, to raise cattle or to fish commercially (Table 8.4). Consequently, most farmers only plant about a quarter to a half of a hectare of bitter manioc, beans and maize for self-provisioning. Some farmers sell part of their manioc, but two-thirds of the typical families of the conservation units earn no income at all from cropping. As a result, most of the residents of the national forest are dependent on government transfer payments in the form of retirement pensions and family support programmes for monetary income.

The attempts to build multifunctional rural livelihoods have clearly been unsuccessful. Lack of opportunity has induced young people to leave the national forest while the elderly who stay behind live off pensions and head

Table 8.4 Land use of small riverine farmers of the Tapajós River

Type of Farmer	Cropping		Pasture for Cattle		In Short or Long Fallow		Forest		Total	
	ha	%	ha	%	ha	%	ha	%	ha	%
Inside Tapajós National Forest	1,0	1,4	0,4	0,5	7,8	10,9	62,5	87,2	71,7	100,0
Outside Tapajós National Forest	0,5	0,6	5,8	6,9	1,6	1,9	75,9	90,6	83,8	100,0

Source: Field research (2008)

48% of the households. Of the children over 18 years old, only 38% still live in the national forest while 58% have left to live in Santarém, Manaus and other regional cities.

These trends demonstrate that the local people in charge of the community-based logging cooperative, when faced with the environmental and logistical limitations outlined earlier, were right to concentrate on extracting timber. Over time, riverine farmers have undertaken many of the non-timber activities promoted by governmental and non-governmental environmental groups but have eventually given them up when they ceased to generate sufficient income.

Latex is a case in point. In recent years, demand has picked up for producing ecologically friendly tires and condoms from natural materials, so latex is being extracted again for the first time in decades. The demand for latex tires for racing cars and condoms for allergic users may not create a significant export market, but the opening of a factory in Manaus in 2012 where latex is used to produce tires for motorcycles and bicycles for this booming regional market is more promising. This notwithstanding, income from latex was still quite modest for the farmers interviewed in 2013, representing 9% of the income of farmers who lived outside the conservation units, 3% for those inside and 4% for cooperative workers. This is insignificant when compared to the income earned from other rural activities and is a shadow of what was made during the rubber boom of the past.

Conclusion: were the Ford Company towns failures?

The Ford company towns, as plantations meant to produce latex on an industrial scale for the global market, were indeed failures. However, the towns left behind are not ghost towns reclaimed by the forest as is popularly thought to be the case. Fordlandia lost much of its urban population. During the 1930s, about 5,000 inhabitants lived there while about 1,000 do today (Geller 2012). Nevertheless, it is the second-largest urban centre of Aveiro municipality where about half the urban population are residents in the county seat. Given the depressed state of the land settlement projects in the vicinity, the rural economy of Fordlandia may not be very dynamic, but the town is strategically located midway between Aveiro and Itaituba and is a regular stop for riverboats that ply the Tapajós. It has schools and clinics that are inferior to those that existed in the past, but they are the best available on this stretch of the river. Paving the BR-163 south from Santarém has reduced the time consumed going to Fordlandia by road to about six hours and if, one day, asphalt reaches the town, it may yet be rejuvenated.

Belterra, on the other hand, was transformed when regional development reached western Pará state. The town was raised to municipal status in 1997 and, in 2010, had a population of 6,863 as compared to a local rural population of 9,466 (IBGE 2010). While Fordlandia remains an isolated town, Belterra is becoming part of the urban sprawl of Santarém. This can be seen at night.When lying off Fordlandia at night, one sees twinkling lights spread out in the town following the urban plan from the time of Ford. From the sky, Fordlandia presents a small ellipsoid shape of light in a sea of unlit countryside and forest. Aveiro has a small round shape, while Belterra has the same ellipsoid shape that Fordlandia has – but one that now connects with the lights spreading out from Santarém (NightEarth 2017).

The urban growth of Belterra has followed the old plantation sector organization so that the house addresses are Sector 1, 2, 3 and so forth. After a neo-liberal turn of federal governments after 1990, the Ministry of Agriculture stopped spraying the rubber trees with fungicide. Most of the trees died, but an occasional survivor is encountered in a backyard. The little city is spread out like Fordlandia and has lots of open space and wooded areas that separate the groups of houses from one another. A sort of central square exists with a bandstand from Ford times, and some low-lying public buildings are situated around it. A few small stores are dispersed around the town, but families of the municipality all stated in interviews that they go to Santarém for shopping, for serious health problems and for many other basic services. Belterra may be a separate municipality in political terms, but it is firmly within the economic and social orbit of Santarém.

References

Aldrich, S, Walker, R and Simmons, C 2012, "Contentious land change in the Amazon's arc of deforestation", *Annals of the Association of American Geographers*, Vol. 102, No. 1, pp. 103–28.

Amorim, AT 2000, *Santarém: Uma Síntese Histórica*, Ulbra, Santarém.

Animazon 2017, "Imigração japonesa", www.animazon.com.br (viewed 29 September 2017).

Bicalho, AMSM 2009, "Capital social na várzea amazônica" in AMSM Bicalho and PCC Gomes (Eds), *Questões Metodológicas e Novas Temátimcas na Pesquisa Geográfica*, Publit, Rio de Janeiro, pp. 93–122.

Bicalho, AMSM 2010, "Reestruturação rural e participação política no entorno de Manaus", in RAO Santos Jr and P Léna (Eds), *Desenvolvimento Sustentável e Sociedades na Amazônia*, Museu Goeldi, Belém, pp. 409–46.

Boserup, E 1965, *The Conditions of Agricultural Growth*, Aldine, Chicago.

Camargo, FC and Guerra, AT 1959, "Região Amazônica" in AT Guerra (Ed), *Grande Região Norte*, IBGE, Rio de Janeiro, pp. 11–6.

Castro, T 1973, "O Brasil e a bacia amazônica", *Revista Brasileira de Geografia*, Vol. 35, No. 2, pp. 2–14.

Cruls, G 1939, "Impressões de uma visita a Companhia Ford Industrial do Brasil no estado do Pará", *Revista Brasileira de Geografia*, Vol. 1, No. 4, pp. 3–25.

Dean, W 1987, *Brazil and the Struggle for Rubber*, Cambridge University, Cambridge.

Food and Agriculture Organization (FAO) 2011, *Protection against South American Leaf Blight of Rubber in Asia and the Pacific Region*, Regional Office for Asia and the Pacific, Bangkok.

Foweraker, J 1981, *The Struggle for Land*, Cambridge University, Cambridge.

158   *Ana Maria de Souza Mello Bicalho and Scott William Hoefle*

Geller, J 2012, "Fordlândia", https://sanguesuorseringais.wordpress.com/2012/01/24? (viewed 29 November, 2017).
GoogleEarth 2017, www.google.com/earth (viewed 12 October 2017).
Grandin, G 2009, *Fordlandia: The Rise and Fall of Henry Ford's Forgotten Jungle City*, Henry Holt, New York.
Hecht SB and Cockburn, A 1990, *The Fate of the Forest: Developers, Destroyers, and Defenders of the Amazon*, University of Chicago, Chicago.
Hecht SB and Cockburn, A 2011, *The Fate of the Forest: Developers, Destroyers, and Defenders of the Amazon*, 2nd ed, University of Chicago, Chicago.
Hine, RV and Faragher, JM 2000, *The American West*, Yale University, New Haven.
Hoefle SW 2006, "Twisting the knife: frontier violence in the Central Amazon", *Journal of Peasant Studies*, Vol. 33, No. 3, pp. 445–78.
Hoefle SW 2012, "Soybeans in the heart of the Amazon?", *Horizons in Geography*, Vol. 81–2, pp. 94–106.
Hoefle SW 2013, "Beyond carbon colonialism: frontier peasant livelihoods, spatial mobility and deforestation in the Brazilian Amazon", *Critique of Anthropology*, Vol. 33, No. 2, pp. 193–213.
Ianni, O 1979, *Colonização e Contra Reforma Agrária na Amazônia*, Vozes, Petrópolis.
IBGE (Instituto Brasileiro de Geografia e Estatística) 1970, "Micro-região 12 (PA)" in *Divisão do Brasil em Micro-regiões Homogêneas*, Rio de Janeiro, pp. 30.
IBGE 2010, *Sinopse do Censo Demográfico*, www.ibge.gov.br/@cidades (viewed 24 October 2017).
IBGE 2017, "Produção agrícola", www.ibge.gov.br/@cidades (viewed 29 January 2018).
Kautsky, K 1988 (1899), *A Questão Agrária*, Brasiliense, São Paulo.
Mathis, E 1998, "Garimpagem de ouro e valorização da Amazônia", *Paper do NAEA* No. 101, UFPA, Belém.
Moran, EF 1981, *Developing the Amazon*, University of Indiana, Bloomington.
NightEarth 2017, www.nightearth.com (viewed 9 September 2017).
Nogueira, D 2017, "Altamira, um retrato do Brasil que mata", *O Globo* 13/12/2017, p. 8.
Oliveira, AU 1999, *A Geografia das Lutas no Campo*, Contexto, São Paulo.
Ruthenberg, H 1980, *Farming Systems in the Tropics*, Oxford University, Oxford.
Schmink, M and Wood, CH 1992, *Contested Frontiers in Amazonia*, Columbia University, New York.
Simmons, C, Walker, R and Arima, E 2007, "Amazonian land war in the south of Pará", *Annals of the Association of American Geographers*, Vol. 86, pp. 567–92.
Simmons, IG 1989, *Changing the Face of the Earth*, Blackwell, Oxford.
Soares, T 1973, "O Brasil e a Bacia Amazônica", *Revista Brasileira de Geografia*, Vol. 35, No. 2, Caderno Especial, pp. 2–14.
Survival International 2000, *Disinherited*, London.
Wilson, G and Burton, RJF 2015, "'Neo-productivist' agriculture: spatio-temporal versus structuralist perspectives", *Journal of Rural Studies*, Vol. 38, pp. 52–64.

9 Physical and social engineering in the Dutch polders: the case of the Noordoostpolder

Tialda Haartsen
Frans Thissen

Introduction

On 14 June 2018, exactly 100 years had passed since the *Zuiderzeewet* (Southern Sea Act) was enacted. In this act, it was agreed that the Zuiderzee (Southern Sea), a large shallow inlet of the North Sea in the centre of the Netherlands, would be enclosed by a barrier dam and would be partially reclaimed at the expense of the Dutch state (Constandse 1976). Between then and now, 1,650 km² of new land has been planned and developed in the context of this so-called *Zuiderzee project*. The project consisted of the drainage of the *Wieringermeerpolder* (1930), the completion of the *Afsluitdijk* (Barrier Dam) in 1932 and the related creation of the sweet water lake *IJsselmeer* (Lake IJssel), the drainage of the *Noordoostpolder* (1942) and the polders of *Oostelijk Flevoland* (1957) and *Zuidelijk Flevoland* (1968). The planned polder *Markerwaard* was never drained.

The Zuiderzee project is an outstanding example of the centuries-old tradition of land reclamation from coastal tidal waters, inland lakes and marshes for which the Netherlands is famous. The Netherlands also has a long history of comprehensive planning and landscape design. In the twentieth century, both traditions culminated in the largest national planning project in the Netherlands. It is not only the size but especially the extremely strong influence of the national government in both the reclamation and the design of this new land that makes the Zuiderzee project extraordinary.

The plans for and implementation of the partial reclamation of the Zuiderzee coincided with the development of the social sciences in the Netherlands. In the interbellum, "emerging scholarly disciplines, such as rural sociology, agricultural economics and spatial planning took a central role in defining problems and solutions and setting the political agenda, thus contributing to a 'scientisation of the social'" (Van de Grift 2017, p. 108). Spatial planning was increasingly preceded by extensive surveys. The Zuiderzee project opened new opportunities for designing a society based on "planning blueprints" that were developed in advance (Van der Cammen et al. 2012). According to Constandse (1960), a government has three instruments for developing a new region: physical planning, colonisation politics (measures to organise the admission of settlers to the area) and cultural policy (measures to influence community life in the area). In the Zuiderzee project, especially in the Noordoostpolder, all three instruments were used to their maximum extent.

Because of the staged development of the Zuiderzee project, the planning process for each polder was evaluated by social scientists and improved, refined and adapted to observed societal changes and implemented in the next polder. In the first two polders, the Wieringermeer and Noordoostpolder, the aim was to create a modern agricultural production area and a rural society based on scientific principles and according to the best traditions of Dutch engineering and planning. In Oostelijk Flevoland and especially Zuidelijk Flevoland, the focus shifted gradually from a predominance of agriculture towards urban development and the creation of natural and recreational areas as well (Van Hulten 1969).

This chapter provides an overview of the Zuiderzee project and specifically describes the design and development of the Noordoostpolder. The Noordoostpolder is chosen as a case study, because modernist, top-down, blueprint planning reached its climax here. As a result, the Noordoostpolder is considered to be the most artificial agricultural landscape in the Netherlands (Figure 9.1). Also, the social environment was designed according to science-based selection procedures of the first

Figure 9.1 Aerial photograph of the landscape of the Noordoostpolder.

Source: Marco van Middelkoop/Aerophoto-Schiphol

inhabitants. The chapter also evaluates and reflects on the design and development of the Noordoostpolder. In doing so, we discuss whether the Dutch planners succeeded in designing an optimal agricultural production area and a well-ordered and aesthetically pleasant social environment, which is what they aimed for. Finally, we conclude the chapter by making a plea that the time is ripe to do more than to reflect. We contend that this very large and extraordinary Zuiderzee project deserves much more (inter) national recognition.

History and context of the Zuiderzee project

The Zuiderzee project fits very well into the history and the image of the Netherlands as a "self-made land", a country shaping and protecting land from the forces of nature, especially the "water wolf" (Van de Ven 2004). The northern and western parts of the Netherlands are flat, low-lying deltas, making the land vulnerable to flooding by rivers and the sea. The Dutch struggle against the sea started with the building of earthen dwelling mounds in the northern Netherlands from 500 BC. Since 1000 AD, dykes have been constructed to protect the land from flooding and to drain and reclaim some of the lost land (Hoeksema 2007). In the sixteenth century, relatively large areas were flooded and lost because of postglacial sea level rise, followed by an increase in land reclamation.

Hoeksema (2007) distinguishes three stages in the history of Dutch land reclamation. The first stage is the drainage and reclamation of several lakes north of Amsterdam, in the sixteenth and seventeenth centuries. Urban merchants set up these projects as a capital investment. Windmills were used to pump the water out of the lakes. This resulted in polders with a rational system of orthogonal parcels of land, of which the *Beemster* polder (7,000 hectares), reclaimed in 1612, is the most famous (Constandse 1972). The second stage was the drainage of Lake Haarlem (or the *Haarlemmermeer*) near Amsterdam, of 18,000 ha in 1852. This differed from the reclamations in the first stage in the large size of the lake, the use of steam power to drain it and the fact that this reclamation was a public works project (Hoeksema 2007; Van de Grift 2013). The third stage was the twentieth century closure and partial reclamation of the Zuiderzee.

Although the first plans for the reclamation of the Zuiderzee were made in the seventeenth century, the technology to achieve it and agreement on the usefulness and necessity of this project did not occur until the end of the nineteenth century. In 1886, the Zuiderzee Association was founded in order to promote the enclosure and reclamation of the Zuiderzee. Around the year 1900, several plans circulated in the Netherlands. One of them was by the civil engineer Cornelis Lely who was commissioned by the Zuiderzee Association (Van Hulten 1969; Van de Ven 2004).

The plans for the Zuiderzee project were highly contested (Hakkenes 2017). Opponents of the project were found in the circles of fisheries and politics.

The proponents, mainly the liberal elite united in the Zuiderzee Association, did not only seek to increase safety but also sought an increase in the amount of agricultural land in order to foster the national economy (Van Hulten 1969). They also raised the argument of restoring the internal cohesion of Dutch society that was supposed to have been deeply divided by "pillarisation", a segmentation of the society along religious and political lines (De Pater 2011). Moreover, they claimed that the project would improve the position of the Netherlands in Europe. In their opinion, the Zuiderzee was a nationalistic project and as such a part of a "national environmental ideology" (De Pater 2011). Although the foundation for such an ideology lay in the "struggle against the water", this was in certain respects an "invented tradition" (Knippenberg 1997). "The struggle against the water did not shape Dutch national identity, but it became a symbol of the Dutch nation" (Knippenberg 1997, p. 38).

Planning projects such as the Zuiderzee project fitted very well into the international *Zeitgeist* of the interwar period (Van de Grift 2015). In Europe, "hydraulic politics" (De Pater 2011), the linking of major hydraulic projects to nationalistic ambitions, was quite common at that time. Many governments initiated environmental projects as symbols of modernisation and unification (Renes and Piastra 2011). Dutch planners were well informed about colonisation projects elsewhere in the world (Takes 1948; Renes and Piastra 2011). Although the Zuiderzee project was presented as "Holland's war with the sea" (Figure 9.2), the Dutch liberal elite wanted to show the world how to expand their national territory in a peaceful way (Knippenberg 1997) by "internal colonisation" (Van de Grift 2013, 2015, 2017). The reclamation of the Zuiderzee therefore became an object of national pride, attractive to the whole nation and a useful symbol of the Dutch national identity.

Two incidents contributed to the final acceptance of the plan. In 1916, a storm flood showed once more how dangerous the Zuiderzee was, while the First World War made clear how important national food production was. It was a proposal by the civil engineer, Lely, that formed the basis for the actual Zuiderzee project (Figure 9.3). Lely, at that time, was Minister of Water Management in the national government. The act for the Zuiderzee project was piloted through parliament by Lely in 1918. The plan was founded on four objectives: the shortening of the coastline by 270 km to increase safety and to decrease the costs of dyke maintenance; the reclamation of fertile land to enlarge the agricultural area and increase national food production; the improvement of water control and the provision of a better connection between the urban west and the rural north of the Netherlands (Constandse et al. 1982). The execution of the plan started in 1927.

The case of the Noordoostpolder

The Noordoostpolder is the second of the four polders that were reclaimed in the Zuiderzee project and the last in which the creation of a modern agricultural production area and rural society was the main, and indeed the only, goal.

Figure 9.2 Pamphlet promoting Dutch engineering as a victory of the Dutch lion over the "arch enemy" (the sea) and as a symbol of national identity.

Source: De Pater 2011, Figure 2, originally published by The Netherlands Abroad, 1930

The principles used in the planning of the Noordoostpolder were the result of a learning process from experiences with older polder projects. Prevailing ideas for the design and colonisation of the first polder, the Wieringermeer, were based on a study on the experiences in the Haarlemmermeer polder by the human geographer Ter Veen (1925). In the Haarlemmermeer, the liberal government restricted its activities to drainage, leaving the development of

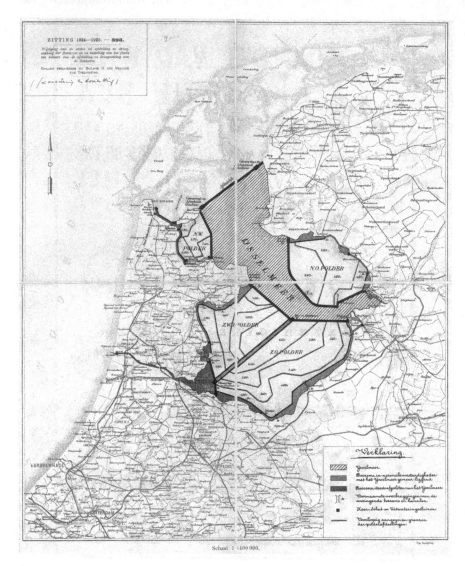

Figure 9.3 Map presenting the enclosure and partial reclamation of the *Zuiderzee* by a barrier dam (*Afsluitdijk*), the creation of a sweet water lake (*IJsselmeer*) and four polders: N.W. Polder (*Wieringermeer*), N.O. Polder (*Noordoostpolder*), Z.O. Polder (*Oostelijk en Zuidelijk Flevoland*) and Z.W. Polder (*Markerwaard*, cancelled). The map relates to an Act of Parliament adopted in 1925, which was a modification and financial reinforcement of the original Act of Parliament adopted in 1918.

Source: *Handelingen en bijlagen van de beide Kamers der Staten-Generaal 1924–1925.* Collection Batavialand, Lelystad

the polder to the market (Constandse 1972). Ter Veen evaluated the colonisation of the Haarlemmermeer polder as a waste of human energy and capital and was very influential in the discussions about the colonisation of the polders in the Zuiderzee project (Heinemeijer et al. 1986).

As in many European state projects in the interwar period (Van de Grift 2017), technological innovation was combined with social renewal. With the introduction of a special type of governance and new forms of top-down technocratic government interventions, "the state is building a new society" (Van de Grift 2017). Two government bodies were established to operationalise this new society: the *Dienst der Zuiderzeewerken* (Office of Zuiderzee Works), which was responsible for the technical aspects of the actual drainage of the polder areas, and the *Directie van de Wieringermeer* (Department of the Wieringermeer), which was responsible for the physical and social engineering of the area (Van Woensel 1999).

The physical and social planning of the Noordoostpolder reached new heights after 1945. Engineers, architects, spatial planners and social scientists became involved in the development of this polder. This period was characterised by a strong belief in progress that was based on an unshakeable confidence in the empirical sciences (Van de Grift 2017). The plans for land allotment, settlement structures and the selection of new inhabitants were important subjects for this type of social research and social planning.

Physical engineering

In order to achieve the goal of "develop(ing) a prosperous and modern agricultural area" (Gort and Van Oostrom 1987), the first point of departure was to create an optimal land parcellation structure. In the Wieringermeer, standard parcels of 20 hectares were used to determine the pattern of farmsteads, roads and canals. A long-lease system was introduced through which the farmers obtained the right to use the land for a fixed price. It was felt that too many farm holdings of the same size (30–60 ha) had been created. This was seen as resulting in a shortage of farm labourers and a lack of a "natural hierarchy" in farm sizes.

In the Noordoostpolder, the land parcels were slightly larger than those in the Wieringermeer polder: 24 hectares. However, more variety in the size of farm holdings was achieved by two contesting developments. "On purely economic grounds, there was strong pressure for an increase in the size of the commercial unit: more machines, fewer people. On social grounds, there was a desire to keep this scale magnification within limits: there was a great demand for farming units, of the traditional type, and the smaller the holdings were made, the more families could be given an independent living. This led to a compromise under which the biggest holdings were to be 48 ha, the smallest 12 ha and the average 24 ha" (Constandse 1976, p. 8). Smaller farms were situated near the villages and larger farms were located further away, so that the majority of the population lived relatively close to the rural service centres.

Because of the previous experiences with parcellation plans and the more or less spontaneous development of settlements in the Haarlemmermeer and Wieringermeer, much more attention was given to the development of an appropriate settlement plan for the Noordoostpolder (Van Hulten 1969). The settlement plan of the Haarlemmermeer only emerged after the first spontaneous housing developments occurred. Ideas about differences in the numbers of inhabitants and functions among settlements were therefore the result of spontaneous development (Takes 1948). More attention was given to the settlement planning of the Wieringermeer. A settlement that was to be the seat of local government was planned in the middle of the polder. For the development of other potential settlements, 13 plots were reserved at distances of 4 km to 5 km from each other. However, villages only developed at three locations, while the originally planned main village proved to be less successful than one of the other villages (Van Hulten 1969).

The first version of the settlement plan of the Noordoostpolder was presented in 1938, two years before the drainage was completed. Changes were proposed in 1939 and during the Second World War, in 1942 and 1943. The final plan was presented in 1946. The debates and research reports about the plans and the subsequent changes were based on a diversity of arguments. However, three topics were central: expectations regarding the number of inhabitants, ideas about the proportion of people living within settlements and in the surrounding countryside (on farms and in small hamlets), and the expected mobility of the inhabitants (Takes 1948; Van Hulten 1969; Constandse 1972, Heinemeijer et al. 1986).

The Noordoostpolder was planned as an agricultural society similar to those of the existing, non-reclaimed rural parts of the country (often referred to as "the old land", in contrast to the new, reclaimed land). The expectations of the numbers of inhabitants were strongly based on the expected agricultural employment levels, which in turn depended on the characteristics of the agricultural structure. Over the period from 1938 to 1946, pressure to accept smaller farms and more labour-intensive crops developed. The final plan also reflected an intensive debate over the best place to house farm labourers: relatively near the farms, and therefore in dispersed dwellings or in hamlets, or living with the nonagricultural population in the villages (Heinemeijer et al. 1986; Van Woensel 1999). The final choice – to concentrate most farm labourers in the villages – was, above all, based on financial considerations. However, houses for the so-called "first labourers" were planned near the farms, in small clusters of two to four dwellings. Assumptions about the mobility of the inhabitants were based on the situation at the time in comparable regions on the mainland.

One of the first decisions was to plan for a single regional centre with approximately 10,000 inhabitants. The remaining settlements were perceived as rural service centres, evenly distributed around the regional centre. Originally five, but later six, such settlements were planned at a distance of 7 km to 8 km from each other, based on the use of bicycles and cars (see Figure 9.4). After conducting surveys in several rural areas on "the old

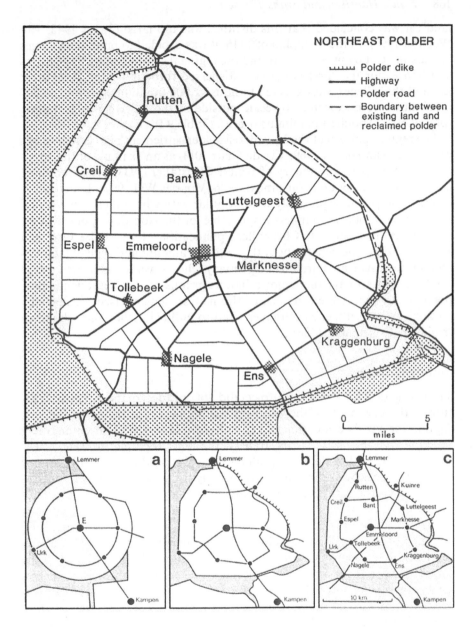

Figure 9.4 The settlements in the Northeast Polder (*Noordoostpolder*): the final settlement plan of 1946 (top) and the development of the settlement plan for
the Noordoostpolder between 1938 and 1946 (a, b, and c).

Sources: De Bruin et al. 1991, Figure 1; Meijer 1981

land", it was concluded that this distance was too large (Takes 1948; Van Woensel 1999). In the final plan of 1946, it was decided to build ten villages at a distance of 6 km centred around the regional centre and with the farms and houses for the "first labourers" dispersed throughout the rural area. Each of the villages was expected to house 2,000 inhabitants, while 20,000 inhabitants were expected to reside outside of the settlements, making a total of 50,000 polder inhabitants overall (Table 9.1).

The regular pattern of the Noordoostpolder settlement plan is sometimes presented (Van Hulten 1969; Van de Grift 2013) as an example of the application of the settlement theory of Walter Christaller (1933). However, there is no direct connection (Constandse 1986; Van der Wal 1997). Dutch geographers and planners probably were familiar with Christaller's theories on central places, but the planners responsible for the Noordoostpolder settlement plans did not refer to Christaller's publications (Constandse 1986; Van Woensel 1999). The debate about the required amount and type of facilities and services, such as churches and primary schools, was complicated by the political wish to have the major religious groups of the Netherlands proportionately represented in each of the new settlements (see "Reflection" section). As a result, each settlement had to build churches and schools in triplicate (Constandse 1986).

Social engineering

The planners of the Zuiderzee project did not only design the physical structure of the new polders, they also set requirements for the new residents. Again, Ter Veen's evaluation of the experiences in the Haarlemmermeer functioned as the basis for this strategy. In the Haarlemmermeer, no attention was paid to the selection of inhabitants, the building of farmhouses or water drainage. According to Ter Veen, only the psychologically strongest, most innovative, rational, sober, materialistic and practical people managed to make a living in the new land of the Haarlemmermeer (Ter Veen 1925). He advised the responsible state organisations to avoid a "natural" selection of the inhabitants and instead to artificially organise this selection by the state to ensure that only the most skilful people would be given the opportunity to make a new living in the ideal societies that were to be created in the Zuiderzee polders. This was seen as a more humane and rational selection process that would increase the success rate of the project.

Ter Veen's advice can be contextualised in the prevailing ideas on Social Darwinism and Eugenics that prevailed internationally in those days (Vriend 2012). This proposed artificial selection can also be referred to as "social engineering": "scientifically founded attempts to govern and influence social behaviour" (Van de Grift 2017 p. 109). Although often associated with totalitarian regimes, social engineering can actually be understood as a characteristic feature of the modern, interventionist, liberal-democratic state as practised in different European countries from the period after the

Table 9.1 Population development of the Noordoostpolder: number of inhabitants of the regional centre, the ten villages and the surrounding countryside of the villages, according to the original plan and the development, 1970–2017

	Number of Inhabitants in Original Settlement Plan	Number of Inhabitants						Indices				
		1970	1981	1987	1995	2004	2017	1970-1981	1981-1987	1987-1995	1995-2004	2004-2017
Emmeloord	10,000	13,129	19,342	20,400	21,830	25,021	25,711	147	105	107	115	103
(as percentage of total population)	20%	41%	52%	54%	55%	55%	55%					
Luttelgeest	2,000	532	687	613	610	690	855	129	89	100	113	124
Marknesse	2,000	1,473	2,161	2,234	2,370	2,880	2,780	147	103	106	122	97
Kraggenburg	2,000	578	699	637	670	800	765	121	91	105	119	96
Ens (incl. Schokland)	2,000	1,410	1,586	1,646	1,770	2,230	2,145	112	104	108	126	96
Nagele	2,000	1,031	1,056	934	960	1,110	1,115	102	88	103	116	100
Tollebeek	2,000	513	571	588	720	1,160	1,850	111	103	122	161	159
Espel	2,000	561	763	684	680	720	820	136	90	99	106	114
Creil	2,000	590	746	637	640	860	1,060	126	85	100	134	123
Rutten	2,000	510	637	574	890	870	925	125	90	155	98	106
Bant	2,000	548	676	631	610	810	750	123	93	97	133	93
Total villages	20,000	7,746	9,582	9,178	9,920	12,130	13,065	124	96	108	122	108
(as percentage of total population)	40%	24%	26%	24%	25%	27%	28%					
(variation coefficient or relative standard deviation of the number of inhabitants in ten villages)	0,00	0.50	0.54	0.62	0.60	0.61	0.54					
Total surrounding countryside villages	20,000	11,129	8,621	8,151	8,190	8,412	7,783	77	95	100	103	93
(as percentage of total population)	40%	35%	23%	22%	21%	18%	17%					
Total Noordoostpolder	50,000	32,004	37,545	37,729	39,940	45,563	46,559	117	100	106	114	102

Source: De Bruin et al. 1991, table 1; Municipality Noordoostpolder; Statistics Netherlands

First World War until the beginning of the 1960s (Couperus et al. 2015). A key eugenic element of the suggested selection was the "biologist paradigm": "the idea that the physical and mental qualities of individual farmers shape the quality of the community" (Van de Grift 2017, p. 125).

In the 1930s, a selection procedure was designed to optimise the colonisation of the Wieringermeer polder (Gort and Van Oostrom 1987; Van de Grift 2017). All those involved in the planning of the polder areas were convinced that the colonisation needed strong leadership (Vriend 2012). A separate section of the *Directie* (the state organisation responsible for the polder's physical and social engineering) was in charge of selection, with director Lindenbergh taking an authoritarian lead. In the Wieringermeer, both future tenant farmers and agricultural workers were selected. The future tenants had to meet both economic and sociobiological selection criteria. The sociobiological criteria encompassed practical professional competence, agricultural education and experience and family composition. Candidates had to be younger than 45 years old, of good health and physically strong. Initially, small business operators in retailing were able to settle freely. The number of applicants for the Wieringermeer polder was not very large because of the economic crisis of the 1930s and the uncertain reputation of the new polder area (unknown and feared) (Gort and Van Oostrom 1987). Despite the relatively low number of applicants to choose from, in 1938, the colonisation procedures for the Wieringermeer were deemed to have been successful. Therefore, they could be used for the Noordoostpolder as well. In agricultural terms, the Wieringermeer polder distinguished itself in a positive way from the "old" agricultural areas. The settlers differed because of their self-confidence, entrepreneurial spirit, workability, endurance, strong vitality and down-to-earth rationalism (Van Heek 1938).

The selection procedure was refined in the post-war years to select new settlers for the Noordoostpolder (Van de Grift 2017). Because of the shortage of houses and the increased demand for agricultural products after the Second World War, the number of applicants for the Noordoostpolder outnumbered the amount of available farms. In the period 1947 to 1957, 1,480 agricultural and 200 horticultural farms were allotted in annual registration rounds. In total, approximately 22,000 applications to become a farmer on the new land were made by around 10,000 applicants. Although most unsuccessful applicants gave up after the first attempt, there was a small group who applied over and over again (Vriend 2012); 5,000 applications were made for agricultural worker positions and at least 2,000 for local retailing opportunities (Gort and Van Oostrom 1987).

Most selection criteria were similar to those for the Wieringermeer, but they were applied more strictly because of the greater popular interest in the Noordoostpolder. Minor tightening of already defined criteria included a narrower age limit (26 to 50 years) and more severe agricultural criteria: candidates needed agricultural knowledge and skills, main work experience in agriculture and the skills to run an agricultural enterprise in a modern and rational way (Gort and Van Oostrom 1987). Furthermore, the criteria for financial capacity changed over the course of the selection period. It began at 1,000 guilders per hectare in 1947 and rose to 1,600 guilders per hectare by 1955.

A new requirement for the Noordoostpolder was that candidates needed to have a "pioneer spirit", which was operationalised as being active in community organisations (e.g., church, school, village councils, sports clubs, agricultural organisations, etc.) and being able to adapt to new situations. This aspect of community development was given a key role in the Noordoostpolder (Van de Grift 2017). In relation to this, the capacities and enthusiasm of wives were also more closely checked in the selection procedure. A man was only thought to be capable of focussing on developing a modern farm and contributing to community development if his wife was very supportive. Therefore, candidate-farmers had to be married or to have plans to get married to a neat and flexible woman who could deal with living in a new land (Gort and Van Oostrom 1987).

For the implementation of the selection of settlers, Lindenbergh appointed and trained seven assessors who were capable of judging people and quickly estimating, from a conversation, whether or not a candidate was suitable for a life on the Noordoostpolder. Each application round was announced through advertisements in newspapers and agricultural magazines. People could get more information by asking for a prospectus and visiting one of the information days held on the polder itself. If they were interested, applicants had to fill out an extensive application form to show how they met all the selection criteria. They also had to include the names of four referees. The selectors requested information from the mayor of their place of residence and from the candidate's agricultural organisation. When all personal files for one allotment round were complete, the first selection took place. After that, the qualified candidates were visited without notice at their homes by one of the selectors. The aim was to see how the candidates functioned in their daily lives. Both the farm and the household were inspected, and the candidate and his wife were interviewed. The information from the home visits was added to their personal files. After that, the final selection took place (Gort and Van Oostrom 1987). According to Vriend (2012), a short list of 5% to 10% of the total number of applicants were actually visited.

During the selection, the first assessment was of whether someone was suited to receive a farm. After that, the size of the farm and its location were determined on the basis of the observed qualities of the applicant and his wife (Gort and Van Oostrom 1987; Van de Grift 2017). Candidate-tenants of a 48-hectare farm had to meet higher requirements (more leadership, real pioneers) than 12-hectare candidates (hard working, docile, committed people).

The most difficult question to answer was whether or not the candidate had a pioneer spirit and was therefore suited to living in a new area where the local community had to be developed from scratch. Lindenbergh explained (Vriend 2012, p. 42): "The technical information, such as age, finances, education, were easy to check. But the suitability of the candidate was the most difficult criterion. .... I had a clear idea of the type of persons I needed. However, it was difficult to judge if people were able to contribute to community development". Nevertheless, and in retrospect, Lindenbergh admitted that the suitability of individual farmers to contribute to community building was, in the end, decisive in their selection (Wolffram 1994).

Reflection

Physical engineering

During the construction period of the settlements in the Noordoostpolder, from 1945 until 1962, it became clear that society, economy and agriculture were all undergoing impressive changes. The most significant was the transformation of the national economy from an agri-industrial to a service-led economy. Although the percentage of agricultural employment was already declining at a national level before the Second World War, the absolute number of people employed in agriculture in the Netherlands increased until 1947 but declined very fast in the 1950s because of mechanisation and rationalisation (Van Leeuwen et al. 2010). Employment growth in the tertiary and quaternary sector and the regionalisation of daily life because of the growing mobility of the Dutch population that characterised post-industrial society in the Netherlands began at the end of the 1960s (Van Engelsdorp Gastelaars and Ostendorf 1986).

For evaluation purposes, and in order to develop and adapt the plans for the next two polders, several studies became available in which the agricultural structure and the settlement plan of the Noordoostpolder were evaluated (Takes 1948; Van Hulten 1969; Constandse 1972; Hoekstra 1980; De Bruin et al. 1991). With regard to the agricultural structure, the average farm size very soon appeared to be too small for mechanised farming, and far fewer agricultural labourers were needed. Because of a shortage of land, most farmers started to intensify their cultivation by growing open field vegetables such as onions and carrots, instead of the traditional crops for which the polder had been planned (potatoes, sugar beets and grain) (Gort and Van Oostrom 1987). Despite this, Hoekstra (1980) concluded that the Noordoostpolder developed, up to the mid-1970s, as a prosperous, relatively independent agricultural region. However, regionalisation, with respect to migration, use of more distant service centres and more extensive commuting and marriage patterns soon became more important in this part of the Netherlands. The regional centre of Emmeloord developed as a service centre for the adjoining municipalities, while the inhabitants of the ten villages developed a stronger orientation towards Emmeloord and also Zwolle, the nearest urban centre on the mainland. At the same time, sociocultural differences between the Noordoostpolder and the mainland diminished. The relatively weak position of the ten villages as rural service centres was also reflected in the lower attraction of these villages as places to live (Hoekstra 1980).

The ten villages showed a less than positive population progression (Table 9.1). Only two villages reached the intended number of 2,000 inhabitants; five never even reached 1,000 inhabitants. Instead, the regional centre of Emmeloord developed fast in absolute and relative terms. As in other rural regions in the Netherlands (Thissen 1995), the development of the settlements was the result of a concentration process caused by age selective migration patterns and a sharp decline in agricultural employment up to the 1980s (De Bruin et al. 1991). As a result, most villages in the Noordoostpolder experienced a decline in population in the 1980s. From the 1980s onwards, population changes in villages in Dutch rural areas became less dependent on local service functions and local employment, while the role of the residential function became more important (De Bruin et al. 1991; Thissen 1995). The relatively positive population development of these villages in the 1990s and the first decade of this century is the result of later ripples in the demographic development of the original, relatively homogeneous, young polder population (Van der Bie et al. 2012) and the popularity of these villages as residential environments during this period. More recently, demographic changes and a new wave of urbanisation (Bijker and Haartsen 2012) have resulted in age-selective outmigration, especially of young adults from the Noordoostpolder (Haartsen and Thissen 2014). It is clear that the planned villages have experienced a more diverse population influx than the planners had foreseen (as indicated by increasing indices of variation up to 1987; see Table 9.1). This can be explained by the sequence of their development (e.g., the earliest village of Marknesse became relatively large) but also by their relative location with reference to commuter and service centres on the mainland and differences in their attractiveness as residential environments.

A study by Takes in 1948 followed the "evaluation tradition" of analysis and sought to guide the design and colonisation of the next polder, Oostelijk Flevoland. At first, this polder was also intended to develop as an agricultural society. However, the focus changed as early as the 1950s (Constandse 1986). This is reflected in the land use designations of the last two polders. Instead of the expansion of agricultural land, other objectives gained in importance. The shortage of urban land for the enlargement of the Randstad-area in an outward direction became apparent (Van Hulten 1969). The number of settlements was strongly reduced in Oostelijk Flevoland, and it was suggested that the final polder, Zuidelijk Flevoland – if it continued to be planned as an agricultural polder – would not need a service centre at all.

Social engineering

In the period after the Second World War, the social engineering that took place in the Noordoostpolder was accepted as natural and not unique. Other examples in the 1930s were the recruitment procedures, including home visits, for potential employees of the Dutch Lamp Factory Philips, and the selection procedures for farmers in villages such as Giethoorn. Of course, in the Noordoostpolder, the *Directie* itself continuously asked social scientists to evaluate whether the selection procedures were working as intended (Vriend 2012; Van de Grift 2017). These evaluations were also aimed at justifying the colonisation process. In 1952, De Blocq van Kuffeler concluded that the selection of both farmers and agricultural workers ensured an optimal yield from the agricultural land. According to him, the criteria that were used in order to achieve the overall goal of the Zuiderzee project were fully justified from a national economic perspective. The new land required people who were competent and suited to making these valuable state investments productive in the interests of the whole nation (Vriend 2012). In 1955, research into the societal roles of farmers in the Wieringermeer and the Noordoostpolder showed that the number and the importance of an individual's societal roles increased with farm size (Vriend 2012). Those selected for the larger farms did indeed take up pioneer roles in community development. So, the selection process had worked.

During the colonisation of the Noordoostpolder, some frictions arose between the *Directie*, which wanted to perform an optimal selection for the benefit of the new rural society, and the national government, which had a broader perspective. At the start of colonisation, the national government had demanded that the population of the Noordoostpolder had to have a similar mix of religion and geographical origin as did the Netherlands as a whole (Gort and Van Oostrom 1987). Also, the approximately 600 to 700 so-called pioneers and *polderworkers* who performed strenuous reclamation work during and after the Second World War had to be favoured. They had been promised farms in prospect but, once the allotment started, the *Directie* unexpectedly also demanded that they fulfil the selection criteria. Although the first two rounds of allotment (about 225 farms) were exclusively reserved for these pioneers, in the end only 54% of the pioneers and 30% of the *polderworkers* were granted a farm (Gort and Van Oostrom 1987; Vriend 2012). Another dispute related to the inhabitants of some communities in the surrounding regions. Their way of existence (fishing) was threatened by the creation of the new land. The national government felt that the *Directie* had to compensate these people by offering them new opportunities in the polder. But the *Directie* claimed that these people did not fit in with the requirements for the envisioned ideal polder society. They were considered to be backward, negative and suffering from psychological decline. The *Directie* refused to give them priority and suggested that they would be better off if they were sent to other parts of the Netherlands (Gort and Van Oostrom 1987).

Over the colonisation period, the national government urged the *Directie* to compensate groups of people who had been disadvantaged by developments in the countryside on "the old land" with a farm on the new land

(Constandse 1976). Examples included farmers who lost their land because of land consolidation or urbanisation and victims of the flooding in the province of Zeeland in 1953. The *Directie* was not in favour of this because these individuals were not always the most rational and innovative famers, and the majority practised the Protestant religion. They would therefore disturb the desired religious balance.

Despite the general acceptance of the selection procedures, some criticism of them arose during the 1950s. These criticisms were especially expressed in the House of Representatives of the national parliament. One example was a motion submitted by Representative Engelbertink in 1951 that demanded more clarity on how the interests of different groups of candidate-tenants were taken into consideration. This move resulted in the installation of a committee of trusted representatives who had to double-check the selection procedures. Interestingly, the Director of the *Directie* became the chair of this committee, so the *Directie* kept control over these deliberations. Because the *Directie* was directly responsible to the relevant ministers, it could function relatively independently and sometimes without reference to wider society (Gort and Van Oostrom 1987).

In the 1960s and 1970s, following sociocultural changes in Dutch society, the selection procedures were criticised more strongly and, in the polders of Oostelijk Flevoland and Zuidelijk Flevoland, they were applied much less strictly. A complete description of the colonisation and selection procedure was provided by Gort and Van Oostrom in 1987. These authors, for the first time, also reflected on the perspectives and experiences of the people who had been selected. Much later, in 2012, Vriend published a historical novel on the selection of people for the new land that also included the painful stories of those that were not selected. Vriend (2012) discussed the selection procedure critically but managed to situate it very well within the timeframe of the interbellum and post-war period, characterised by the development of a strong welfare state and the need to provide housing, land and jobs in the context of a fast-growing population. Her conclusion was that the selection procedure was a suitable instrument that functioned well in this period of rational, modernist planning.

Conclusion

The Zuiderzee project is the largest and most extraordinary of all the national planning projects in the Netherlands. It is not only the size but especially the extremely strong influence of the national government in both the reclamation and the design of this new land that makes the Zuiderzee project special. It is an excellent example of the type of modernist blueprint planning that was popular from the 1930s to the end of the 1960s in that it followed the principle of "survey before planning" in order to create ideal societies (Van der Cammen et al. 2012). In the Noordoostpolder, this blueprint form of planning was at its peak. However, in retrospect, some scholars now claim that the total Zuiderzee project can be considered a major example of adaptive process planning *avant la lettre* because, in each new polder, the plans were adapted to new circumstances (Constandse 1976; Van der Cammen et al. 2012). Following this line of reasoning, the Zuiderzee project was even more innovative given that adaptive process planning only developed during the 1970s.

In 2011, the Noordoostpolder was nominated to become a UNESCO World Heritage site because of its unique planning history. Despite the enthusiasm of the mayor of the municipality, hardly any local enthusiasm was apparent for this nomination. The local population was afraid that heritage status would limit the entrepreneurial spirit in the area. Given the very special history of the Noordoostpolder, we think that the time is ripe to restart such discussions. The year 2018 marked the 100th anniversary of the Zuiderzee project. Given this milestone, it seems appropriate to ask for more attention to be paid to the special story of the colonisation of the Noordoostpolder. Can it (still) be demonstrated that the state managed to create adaptive, strong communities through strict selection procedures? How entrepreneurial are the descendants of the selected pioneers? Are they aware that their ancestors were pioneers, and (how) do they identify with that? It would be a nice challenge to develop a UNESCO World Heritage Site that takes into consideration the pioneer and entrepreneurial spirit that forms the heart of the regional identity of the Noordoostpolder (Simon et al. 2009). In doing so, such a location may need to be more dynamically maintained than other UNESCO sites. This would acknowledge the extraordinary character of the Noordoostpolder and the Zuiderzee project and would give the project the (inter)national recognition that it deserves.

References

Bijker, RA and Haartsen, T 2012 "More than counter-urbanisation: migration to popular and less-popular rural areas in the Netherlands", *Population Space and Place*, Vol. 18, No. 5, pp. 643–57.

Christaller, W 1933, *Die zentralen Orte in Süddeutschland. Eine ökonomisch-geographische Untersuchung über die Gesetzmäsigkeit der Verbreitung und Entwicklung der Siedlungen mit städtische Funktionen*, Gustav Fischer, Jena.

Constandse, AK 1960, *Het dorp in de IJsselmeerpolders. Sociologische beschouwingen over de plattelandscultuur en haar implicaties voor de planologie van de droog te leggen IJsselmeerpolders*, Tjeenk Willink, Zwolle.

Constandse, AK 1972, "The IJsselmeerpolders, an old project with new functions", *Tijdschrift voor Economische en Sociale Geografie*, Vol. 63, No. 2, pp. 200–10.

Constandse, AK 1976, *Planning and Creation of an Environment a Reappraisal: Experiences in the IJsselmeerpolders*, Rijksdienst voor de IJsselmeerpolders, Lelystad.

Constandse, AK 1986, "Van dorp tot dorp; beschouwingen over zestig jaar planning van nieuwe nederzettingen in de IJsselmeerpolders", in WF Heinemeijer, HJ Heeren, CD Saal, and GHL Tiesinga (red.), *50 jaar actief achter de Afsluitdijk;*

Jubileumbundel ter gelegenheid van het vijftigjarig bestaan van de Stichting voor het Bevolkingsonderzoek in de Drooggelegde Zuiderzeepolders 1936–1986, De Walburg Pers, Zutphen, pp 69–84.

Constandse, AK, De Jong, J and Pinkers, MJHP 1982, The IJsselmeerpolders, RIJP, Lelystad.

Couperus, S, Van de Grift, L and Lagendijk, V 2015, "Experimental spaces: a decentred approach to planning in high modernity", *Journal of Modern European History*, Vol. 13, No. 4, pp. 475–9.

De Blocq van Kuffeler, VJP 1952, "Enkele nationaal-economische problemen bij de kolonisatie der IJsselmeerpolders", *Economisch-Statistische Berichten*, Vol. 37, No. 1844, p. 733.

De Bruin, I, Huigen, P and Volkers, K 1991, "Dorpen onder druk: Nederzettingenproblematiek in de Noordoostpolder", *Geografisch Tijdschrift*, Vol. XXV, No. 2, pp. 161–72.

De Pater, B 2011, "Conflicting images of the Zuiderzee around 1900: nation-building and the struggle against water", *Journal of Historical Geography*, Vol. 37, No.1, pp. 82–94.

Gort, M and Van Oostrom, A 1987, *Uitverkoren: De kolonisatie van de Noordoostpolder 1940–1960*, Waanders, Zwolle.

Haartsen, T and Thissen, F 2014, "The success–failure dichotomy revisited: young adults' motives to return to their rural home region", *Children's Geographies*, Vol. 12, No. 1, pp. 87–101.

Hakkenes, E 2017, *Polderkoorts: hoe de Zuiderzee verdween*, Thomas Rap, Amsterdam.

Heinemeijer, WF, Heeren, HJ, Saal, CD and Tiesinga, GHL (Ed.) 1986, *50 jaar actief achter de Afsluitdijk; Jubileumbundel ter gelegenheid van het vijftigjarig bestaan van de Stichting voor het Bevolkingsonderzoek in de Drooggelegde Zuiderzeepolders 1936–1986*, De Walburg Pers, Zutphen.

Hoeksema, RJ 2007, "Three stages in the history of land reclamation in the Netherlands", *Irrigation and drainage*, Vol. 56, No. S1, pp. 113–26.

Hoekstra, C 1980, *De Noordoostpolder en zijn rand: een geografische en sociologische studie naar regionale interactiepatronen*, Stichting voor het Bevolkingsonderzoek in de Drooggelegde Zuiderzeepolders, Amsterdam.

Knippenberg, H 1997, "Dutch nation-building: a struggle against the water?" *GeoJournal*, Vol. 43, No. 2, pp. 127–40.

Meijer, H 1981 *Zuiderzee/Lake IJssel*, IDG, Utrecht.

Renes, H, and Piastra, S 2011, "Polders and politics: new agricultural landscapes in Italian and Dutch wetlands, 1920s to 1950s", *Landscape*, Vol. 12, No. 1, pp. 24–41.

Simon, C, Huigen, P and Groote, P 2009, "Analysing regional identities in the Netherlands", *Tijdschrift voor Economische en Sociale Geografie*, Vol. 101, No. 4, pp. 409–21.

Takes, CAP 1948, *Bevolkingscentra in het oude en het nieuwe land*, Samsom, Alphen aan den Rijn.

Ter Veen, HN 1925, *De Haarlemmermeer als kolonisatiegebied*, Noordhoff, Groningen.

Thissen, F 1995, *Bewoners en nederzettingen in Zeeland: op weg naar een nieuwe verscheidenheid*. KNAG/FRW UvA, Utrecht/Amsterdam.

Van de Grift, L 2013, "On new land a new society: internal colonisation in the Netherlands, 1918–1940", *Contemporary European History*, Vol. 22, No. 4, pp. 609–26.

Van de Grift, L 2015, "Introduction: theories and practices of internal colonization. The cultivation of lands and people in the age of modern territoriality",
International Journal for History, Culture and Modernity, Vol. 3, No. 2, pp. 139–58.

Van de Grift, L 2017, "Community building and expert involvement with reclaimed
lands in the Netherlands, 1930s–1950s", in S Couperus and H Kaal (Eds.), *(Re)
Constructing Communities in Europe, 1918–1968. Senses of Belonging Below, Beyond
and Within the Nation-State*, Routledge, New York, pp. 108–29.

Van de Ven, GP (Ed.) 2004, *Man-Made Lowlands; History of Water Management and
Land Reclamation in the Netherlands*, 4th rev. ed., Matrijs, Utrecht.

Van der Bie, R, Latten, J, Hoekstra, L and Van Meerendonk, B 2012, *Babyboomers in
the Netherlands: What the Statistics Say*, Statistics Netherlands, The Hague.

Van der Cammen, H, De Klerk, L, Dekker, G and Witsen, PP 2012, *The Selfmade
Land: Culture and Evolution of Urban and Regional Planning in the Netherlands*,
Spectrum/Uitgeverij Unieboek, Houten.

Van der Wal, C 1997, *In Praise of Common Sense: Planning the Ordinary. A Physical
History of the New Towns in the IJsselmeerpolders*, 010 Publishers, Rotterdam.

Van Engelsdorp Gastelaars, R and Ostendorf, W 1986, "The Dutch urban system in
transition: from a productive to a consumptive order?", in JG Borchert, LS Bourne,
R Sinclair (Eds.), *Urban Systems in Transition*, KNAG/FRW UvA, Utrecht/
Amsterdam, pp. 231–42.

Van Heek, F 1938, *Economische en Sociale problemen van de Wieringermeer; een studie
van een kolonisatiegebied in wording*, Stichting voor het Bevolkingsonderzoek in de
Drooggelegde Zuiderzeepolders, Samson, Alphen aan den Rijn.

Van Hulten, MHM 1969, "Plan and reality in the IJsselmeerpolders", *Tijdschrift voor
Economische en Sociale Geografie*, Vol. 60, No. 2, pp. 67–76.

Van Leeuwen, ES, Terluin, IJ, and Strijker, D 2010, "Regional concentration and spe-
cialisation in agricultural activities in EU-9 regions, 1950–2000", *European Spatial
Research and Policy*, Vol. 17, No. 1, pp. 23–39.

Van Woensel, JTWH 1999, *Nieuwe dorpen op nieuw land; inrichting van de dorpen
in Wieringermeer, Noordoostpolder, Oostelijk en Zuidelijk Flevoland*, Stichting
Uitgeverij de Twaalfde Provincie, Lelystad.

Vriend, E 2012, *Het nieuwe land: Het verhaal van een polder die perfect moest zijn*,
Balans, Amsterdam.

Wolffram, DJ 1994, *Interview with Lindenbergh*, www.flevolandsgeheugen.nl/5340/nl/
de-selectie-van-pachters-in-de-nop (viewed 24 January 2018).

10 The elusive dream of "development" on the agricultural land settlement schemes in Papua New Guinea

Gina Koczberski
George Curry
Steven Nake

Introduction

This chapter focuses on the agricultural land settlement schemes (LSSs) established in the 1960s on the north coast of West New Britain Province (WNBP), Papua New Guinea (PNG) (Figure 10.1). This area, with its sparse population, fertile volcanic soils and suitable climate was considered ideal for the development of LSSs. More than 35,000 ha of customary land were acquired by the Australian colonial administration and converted to state agricultural leasehold land for plantation estates and smallholder oil palm LSSs[1]. Settlers, predominantly from mainland PNG, acquired land holdings of approximately 6 ha with individual titles on 99-year agricultural leases. After the success of the Malaysian Federal Land Development Authority (FELDA) scheme (Fold 2000; Sutton 2001; Cramb and Curry 2012), agricultural settlement schemes became a favoured development instrument of the Australian administration in PNG. This was because they were viewed as vehicles to promote export-driven agricultural and economic development. They were also perceived to be a mechanism for transforming indigenous agricultural systems and customary land tenure based on group and clan ownership by integrating Papua New Guinean settlers into the market economy. Thus, as in other developing countries at the time, PNG adopted land settlement programmes to modernise the agricultural economy.

Village farming families from other parts of PNG settled on the LSSs and, like the Australian administrators, they initially associated the scheme with transformative powers to bring them the wealth and "development" that Europeans had enjoyed and that had bypassed them in their remote home villages. At the time of the inception of these LSSs, both the colonial administration and the smallholder settlers held high expectations of the benefits that the LSSs would deliver. Fifty years on, oil palm has become the most important export cash crop and dominates the rural economy of the northeast coast of WNBP. As a result of the oil palm development, the province is now among the most economically prosperous in the country, and smallholder incomes compare very favourably with those earned by

Figure 10.1 Papua New Guinea and study sites.

Source: The authors

cocoa and coffee smallholder farmers. There are now approximately 11,500 smallholder oil palm households in WNBP.

What was not imagined by both settlers and the Australian administrators at the time of the schemes' inception were the demographic, economic and land pressures that would emerge over time to challenge their long-term social and economic sustainability. Based on fieldwork conducted among settlers on the Hoskins and Bialla LSSs in WNBP since 2000, this chapter examines the trials and tribulations of this project of economic and social transformation initiated by the Australian administration and eagerly adopted by first- and second-generation settlers. The chapter traces the initial successes of the LSS through to more recent times and examines the range of factors that have conspired to thwart the realisation of the dream that the settlers held of the LSS delivering the "good life" or what they imagined to be a Western or European lifestyle. We argue that, despite the early sense of lost dreams and their feelings of apprehension about the future, few of the original settlers regret their decision to migrate. Now, second- and third-generation settlers are busy pursuing strategies to enhance their livelihoods and social well-being.

Land settlement schemes as instruments of development

The first nucleus-estate smallholder oil palm land settlement scheme in PNG was established at Hoskins in February 1967 following an agreement between the Australian administration and the British plantation company, Harrisons and Crosfield (Fleming 1972). Under the agreement, provisions were made for the nucleus plantation and milling company to be a joint venture between the administration and the Harrisons and Crosfield (London) group of companies (Fleming 1972). The two parties registered New Britain Palm Oil Development Pty Ltd (NBPOD) as a joint venture company, and the company was granted an agricultural lease of more than 2,185 hectares at Nahavio for the development of an oil palm plantation and a processing mill (Longayroux 1972). The granting of the agricultural lease was conditional upon NBPOD purchasing and processing smallholders' crops and the construction of a mill. Adjacent to the estate, the administration agreed to develop smallholder plantings on an LSS and provide the necessary social services and infrastructure.

The establishment of the LSS and the production of oil palm first attracted the interest of the Australian administration in PNG in the early 1960s following an invitation by the Australian government to the International Bank for Reconstruction and Development (IBRD) (later known as the World Bank) to send a mission to PNG to undertake an economic review of the Territory. The principal aim of the review was to make recommendations to assist the Australian government in planning a development programme designed to expand and stimulate the economy of PNG, thereby raising the living standards of the people (IBRD 1965, cited in Densley 1980).

Among the IBRD's recommendations were the expansion of cash cropping and the introduction of oil palm on a nucleus estate-smallholder model to WNBP in order to diversify the agricultural economy and increase the export income of PNG (IBRD 1965, cited in Densley 1980; Grieve 1986). The proposed nucleus estate-smallholder model involved the establishment of plantation estates with centralised milling facilities to process crops from plantations and surrounding LSS smallholders. Such recommendations endorsed the Australian administration's existing policy on rural resettlement. The rural resettlement policy promoted opening up "new" land for the voluntary resettlement of rural people into "underpopulated" and "unused" areas of PNG to stimulate agricultural and economic development. Although the policy was first implemented in 1952, it was not until the late 1960s that the Australian administration embarked on an ambitious programme of land settlement schemes. Paul Hasluck, the Australian Minister for External Territories at the time, talked of large resettlement programmes in the coastal areas of PNG (Hulme 1984, p. 86). From 1960 to 1967, cocoa, coconut and rubber were the preferred cash crops for the settlement schemes but, following the IBRD's mission and recommendation of oil palm for PNG on a nucleus estate-smallholder model, oil palm became the major crop of the settlement schemes (Hulme 1984, p. 92). In WNBP and Oro Province, existing settlement blocks planted to cocoa and coconut were converted to oil palm.

The nucleus estate-smallholder model proposed by the IBRD encouraged the administration, including Papua New Guinean parliamentarians[2], to expand rural settlement schemes in PNG but to do so on a larger scale than earlier schemes (Hulme 1984). In 1967 and again in 1971, parliamentary missions from PNG toured the Malaysian FELDA settlement schemes to gather information to guide the development and management of existing and future settlement schemes in PNG. The report submitted from the 1971 tour of the FELDA schemes concluded that:

> Large-scale smallholder development represents the most effective use of capital and human resources to bring indigenous farmers into the cash economy. They [LSSs] should be given increasing emphasis.
>
> (SCLD 1971 p. 57, cited in Hulme 1984, p. 95)

In another report, the Commission of Inquiry into Land Matters (CILM 1973) investigated land issues in PNG, and land settlement schemes were examined for their achievements and to consider whether any further schemes should be developed. Among its four recommendations, the report encouraged the resettlement of people from land-short to less populated areas (within their own districts or adjacent districts) and supported the expansion of large-scale land settlement schemes based on the nucleus estate model (CILM 1973, cited in Hulme 1984, p. 99). Land settlement schemes were considered to be key tools of national development.

As instruments of national development, the LSSs were thought to have transformative powers. They were viewed as a means to transform indigenous agricultural systems, increase agricultural export production and rural incomes and to integrate Papua New Guineans into the cash economy. The schemes were also seen as avenues to relieve population pressure in some rural areas and to bring underexploited land into production in other areas (Hulme 1984; Connell 1997; Koczberski and Curry 2005). The Australian administration envisaged that, by establishing individualised holdings on land settlement schemes, the perceived problems of traditional communal land tenure in hindering agricultural development would be overcome. It was thought that Papua New Guineans would soon recognise the benefits of an individualised land tenure system, a recognition that would hasten the replacement of customary land tenure based on group ownership with individual land titles. As Hulme noted in regard to the LSSs:

> A recurrent theme in the observations of colonial administrative officers and agriculturalists was that traditional systems of tenure, based on group ownership, were a major disincentive to agricultural development and that a system allowing for the individualisation and registration of ownership, similar to that in Australia, should be instituted. Such a system, it was (and is) argued, would facilitate rural credit, allow for the transfer of land from groups with a surplus to those which

are land-short and permit energetic farmer-entrepreneurs to maximise their productive capacity.

(Hulme 1984, p. 70)

Thus, the colonial administrators considered it valuable for Papua New Guineans to break with their traditions and lifestyles so that subsistence agriculture would give way to export cash crop production and an individualised system of land tenure would displace indigenous forms of land tenure. In this imagined future, export cash crop production would provide the impetus for regional growth and development and the LSSs would provide land-short migrants with a more secure future. It was within this optimistic framework for agricultural and economic development that the oil palm LSS at Hoskins was conceived.

From risky business to "international showpiece"

Whilst the Australian administration held optimistic visions of the future, the risks involved in such an ambitious project did not go unnoticed by the company. Tom Fleming, who was initially sent by Harrisons and Crosfield to investigate the potential for oil palm development in WNBP, noted in 1972:

> The project involved introducing a new crop into a new and completely undeveloped territory; and it was to be combined with a large-scale social and economic experiment involving an extensive resettlement of people of whom few had previously grown cash crops on a large scale and some would be growing them for the first time. ... Development of the plantation would also involve training completely inexperienced labour and personnel from elsewhere in Papua New Guinea. ... Furthermore, the risks involved were considerable, particularly as the building of a mill much larger than was needed for the plantation crop involved a heavy outlay on which there could be very little return if the smallholdings were not a success.

(Fleming 1972, p. 10)

It was clear from the project's inception that its success would be determined largely by the performance of the newly settled smallholder producers on the LSS.

In July 1968, the first smallholders settled on their 6-ha leasehold blocks. A condition of the lease was that 4 ha of oil palm were to be planted within two years of settling on the block. The remaining 2 ha were viewed as "reserve land" for food gardens, and settlers were expected to be self-sufficient in food from their own gardens shortly after moving onto their blocks. Nearly all leasehold blocks were acquired by people from mainland PNG; preference was given to applicants from land-short areas such as parts of Simbu

Province, Maprik (East Sepik Province – ESP), Wabag (Enga Province) and the Gazelle peninsula of East New Britain Province. Publicity committees from the National Land Board were sent to these land-short areas to encourage people to apply for settlement blocks (Hulme 1984). Many industrious migrant labourers employed in presettlement logging activities were also allocated blocks, and later, Morobe Province was given preferential consideration as a recruitment site (Ploeg 1972; Koczberski et al. 2001). Successful applicants were provided with a loan of up to $AUS1870 from the PNG Development Bank to cover relocation costs, house building, oil palm seedlings, tools, pest and disease control, land rent and a living allowance while waiting for the first harvest (Jonas 1972; Landell Mills Ltd. 1991). By 1980, 11,824 people had settled on the Hoskins LSS (National Statistical Office of PNG 1981, cited in Hulme 1984, p. 246).

Smallholders adjusted exceptionally well to their new circumstances, exceeding both administrator and company expectations. During the development phase, nearly all smallholders attained expected planting targets and production levels, and yields far exceeded those projected (Ploeg 1972; Hulme 1984). From 1971/72 to 1976/77 smallholder production rates were consistently 50% or more above projected production levels, and yields averaged 18.68 tonnes/ha, exceeding 75% of company estate yields. The outstanding production levels achieved by smallholders allayed initial anxieties regarding the potential performance of the new settlers. A few years after the first settlers planted their blocks, Fleming noted retrospectively that "naturally, there were doubts as to how far a large number of inexperienced smallholders would meet the necessary requirements [of meeting harvesting deadlines and time schedules for collection by the company]. In fact they adjusted very quickly and results have been good" (Fleming 1972, p. 19).

Those involved in the development and management of the Hoskins scheme regarded it as "an international showpiece for agricultural development" (Hulme 1984, p. 245). Its perceived success led the government to set up a similar oil palm nucleus estate-smallholder scheme further east along the coast in the Bialla District, and attention also turned to encouraging customary landowners to become involved in oil palm production under the Village Oil Palm Project (VOP) scheme. In 1986, an Asian Development Bank loan funded an expansion of the Bialla LSS and, in 1990, two new LSS subdivisions (Soi and Kabaiya) were established totalling more than 600 blocks. In 1980, a VOP scheme was initiated and, by 1986, 244 VOP blocks had been planted (Koczberski and Curry 2003). At the end of 1994, 7,365 ha of oil palm had been established on the Bialla LSS by 2,306 smallholders (Diapo 1995).

Migration and imagined futures

Many migrants who settled on the schemes in the initial development phase saw themselves as intimately involved with bringing "progress" and "development" to WNBP and to the (from 1975) independent nation of PNG (Koczberski and Curry 2004). Such a claim has compelling evidence. In little over a decade following the establishment of the Hoskins LSS, the north coast of WNBP was transformed from an isolated, sparsely populated, poorly serviced and economically disadvantaged region to one where roads, industrial complexes, social services and urbanisation had expanded (C. Campbell, pers. comm., 1 September 2006). A large wharf, two high schools, the expansion of Kimbe Hospital and new commercial outlets were some of the physical signs of "progress". The regional economy underwent a profound change as money made its way to the province for investment and infrastructure development and income from oil palm smallholders and plantation employees circulated through the local economy. Demographically, the area was also dramatically altered. Not only had the population increased significantly, but the ethnic mix of the population was transformed. Settlers, notably from the provinces of East and West Sepik, Simbu, East New Britain and Morobe, became concentrated in large numbers in the area. By the 1980s, Kimbe rose to prominence as a political and administrative centre when the town was nominated as the administrative headquarters for the province and the seat for the provincial government (Hulme 1984).

Settlers not only viewed themselves as crucial players in the progress and development of the province but, like the colonial administration, they also considered that the LSSs were imbued with transformative powers and held high expectations of the future possibilities that the schemes would deliver in advancing their standard of living. In interviews with settlers, when discussing why they migrated to WNBP, many referred to the perceived opportunities that the LSSs offered and contrasted these to the constraints and problems existing at the time in their home villages. Unlike their home villages, the schemes were viewed as providing a path to a better life for themselves and their children by facilitating access to land for cash cropping, an escape from sorcery and jealousies (which were linked to tribal war in the highlands), good access to schools and healthcare and the opportunity to participate in the market economy and earn a regular income. In one conversation with a settler from the ESP, ongoing disputes over land and his limited access to land and sago in his village were key reasons why he applied for an LSS block at Bialla. By gaining access to land in WNBP, he believed his children would have a better opportunity to earn an income and avoid the threat of sorcery. Kean also noted that, among Tolai men living at Siki subdivision in the Hoskins LSS, the opportunity to settle on the scheme was seen as providing economic benefits and a means to "escape the constant demands placed on villagers to contribute to ceremonial occasions, provide food for distant kin and help finance relatives' business ventures" (Kean 2000, p. 162). Thus, people were attracted to the schemes because they were perceived as offering wealthier, healthier and more educated lives alongside a more autonomous social life away from ongoing home village grievances and the demands of kinship obligations.

However, it was the potential economic transformation of their lives that created a strong attraction and desire among potential settlers. As noted previously, settlers were keen to actively embrace development and, importantly, to experience the economic riches that hitherto had eluded them in their home villages. According to Ploeg, settlers believed that, when they settled their blocks, the returns from oil palm would be sufficient for them to "live like Europeans" (Ploeg 1972, p. 106) – the European world being associated with prestige, wealth, luxury goods and a lifestyle in sharp contrast to their own. This world was greatly desired. As Ploeg wrote:

> The settlers felt that the arrival of the Europeans and their rule in Papua New Guinea had started a radical but welcome transformation of indigenous society which was still far from being completed. The Europeans were still most prominent in this new society and only they knew how to derive economic benefits from it. The settlers felt that Papua New Guineans would gradually acquire the knowledge and status to which they were entitled. In their view an important, if not the most important, task of the Administration was to help them become full members of this new society. In their view many Europeans gained their affluence through managing a *bisnis* and the settlers wanted to imitate them. Because being a *bisnisman* showed that the operator had gained insight into the workings of the new, European-inspired society, running a *bisnis* was prestigious and as such had become an end in itself.
>
> (Ploeg 1972, p. 106)

To the settlers, their settlement block was their *bisnis*[3] and the basis from which further *bisnis* activities could be initiated to provide the path to European-like progress and prosperity. The high expectations that their newly acquired block would deliver such wealth were also likely stimulated by the early outstanding smallholder production levels and a post-war peak in world palm oil prices (Fleming 1972) resulting in very high initial returns to smallholders. In 1978/79, average income per settler from agricultural activities was between 2,000 and 3,000 kina (US\$1 = PNGK0.7292 in 1978). In 1979, per capita income in PNG was estimated at K440 (Hulme 1983, p. 332; Grieve 1986, p. 74).

However, as Fleming observed, the exceptionally good incomes "raised expectations to levels which could not be maintained" (1972, p. 19). When oil palm prices fell from K49.97/tonne in 1977 to K1.63 in 1981 (Adamson et al. 1984, p. 124), and remained low throughout the early 1980s, dissatisfaction and despondency soon crept in as more modest incomes became the norm and the imagined grand prosperity of living "like Europeans" did not eventuate. More frugal spending patterns were observed, especially regarding the consumption of beer, which, as a commodity, was closely associated with a European lifestyle and an item of ceremonial exchange (see Ploeg 1972). As Hulme (1984, p. 286) commented in relation to the fall in prices in the early 1980s, "Most settlers reduced their [beer] drinking to more modest levels, but talked wistfully of the *gut taim tru* (really good time) of the 1970s".

Thus, by the mid-1980s, not only did the European lifestyle prove to be unattainable, but smallholders had their first taste of the vagaries of world commodity prices. It would be another 30 years before smallholders again experienced record high prices for their crop.

Fifty years on: contemporary life on the LSSs

On 10 December 2017, New Britain Palm Oil Limited (NBPOL) celebrated its 50th anniversary at Hoskins. At the 40th anniversary, the celebrations had been reported by Papua New Guinea's national newspaper, the *Post Courier*, as follows:

> NBPOL's 40th anniversary celebrations ran for a full week and Independence on September 17 at the Mosa oval, starting at eight, [and continued] … into the night. There was an afternoon drizzle but the interest of the people proved too strong. Company officials presented long serving employees with commemorative medals and the well-kept oval with adjoining company golf course erupted with the sound of stamping feet, rustle of *bilas* and rattles, pounding *kundus* [drums], clattering bamboos and spirited singing. *Singsing* groups arrived by the truckloads, thundering their way through the company premises. Many composed special songs to honour NBPOL. All PNG provinces were represented, many of them [by] company workers, with the local people performing unique dances. The columns of people and the sea of umbrellas [was] awesome; it was their day and the people made very good use of it. There were camp fires at night in front of staff houses and people dressed in traditional costumes, *kundus* beating and singing going on everywhere.
>
> (*Post Courier* 2007, p. 4)

The celebrations were a time for reflecting on shared achievements. The general manager of NBPOL, when addressing the crowd under the banner theme "Grow with Us", explained that the theme was not to be interpreted as "you are growing because of the company" (*Post Courier* 2007, p. 4). Rather, "the success of the company has come about through your participation so it's your success; you are lifting yourself and your place" (*Post Courier* 2007, p. 4). For settlers who had long seen themselves as key drivers of the economic and political changes of the 1970s and 1980s that transformed WNBP into one of the more prosperous provinces in PNG, these were honorific words of recognition of their achievements.

At the national level, the oil palm growing areas in WNBP rank highly in terms of income and services relative to other rural areas of PNG, which, in part, explains why the province has the second-highest net inward migration rate after the National Capital District. Oil palm is the only export commodity tree crop over the past two decades to show continual growth in PNG (Papua New Guinea Palm Oil Council 2012). A typical LSS block at Hoskins in 2016 with 6 ha of oil palm generated an estimated total annual income of K21,104 (AUS$9,044) (PNGOPRA 2017)[4]. Since 1997, each block has received two payments per month from the milling company: the Papa Card payment for fresh fruit bunches (FFB) paid to the leaseholder and the Mama Card payment for the collection by women of "loose fruit" (ripe fruitlets dislodged from the main bunch during harvesting) and usually paid to the wife of the male leaseholder (Koczberski and Curry 2005). Approximately 23% of this income was earned directly by women. Thus, the regular flow of money into the LSSs on paydays is substantial relative to that of many other rural communities relying on commodity crops.

Similarly, educational levels among smallholders, although still low at 7 years of schooling, are well above the national average of 3.9 years (Ryan et al. 2017). Ownership rates of household assets and new technologies such as mobile phones and DVD players are also relatively high among smallholder households.

Lost dreams and fading prosperity

However, despite four to five decades of regular income flowing into the households on the LSSs at Hoskins and Bialla, there are few signs of material prosperity and savings levels are low. Based on observations and conversations with smallholders and industry personnel and written documentation of the initial establishment phase of the schemes (e.g. Ploeg 1972; Hulme 1984; Diapo 1995), living conditions for many have not improved substantially, and for some, have even deteriorated over time.

As noted earlier, when the Hoskins and Bialla LSSs were established, settlers were provided with a basic two-room timber house and a rainwater tank (Figure 10.2). A walk through the subdivisions today reveals that nearly all the original timber houses are gone. Most people live in semipermanent houses or houses constructed from bush materials (Table 10.1). In addition, many of the original rainwater tanks rusted out years ago, and there is no access to piped water in most of the LSS subdivisions (Figure 10.2). In 2007, many women were obtaining drinking water from open-topped 200-litre drums that caught rainwater or from local streams (Curry et al. 2007). Only 18% and 19% of LSS blocks at Bialla and Hoskins, respectively, had rainwater tanks.

Figure 10.2 Original settler house in foreground, bush material house in background and 44-gallon drums used as water tanks.

Source: The authors

Other basic services, such as health services and infrastructure, have also
declined. This is in line with the general trend in PNG of declining government services and funding shortages for local level government services
(Pincock 2006; UNDP 2014; World Health Organisation 2015; Reilly et al.
2015). Over the years, the aid posts on the LSSs have closed or have operated
with shortages of medical supplies. Government funding of agricultural
extension services to oil palm farmers has also decreased, resulting in severe
financial and institutional constraints and falling staff levels. The extension
officer to farmer ratio at Bialla increased from 1:42 in 1982 (Hulme, 1984)
to 1:136 in 1994 (Diapo 1995) and to more than 1:300 growers in 2012 (Ryan
et al. 2013).

The paradoxical situation of regular and relatively high monthly incomes
over many decades of oil palm production and poor, if not declining, living
standards is explained by a mosaic of micro- and macro-level institutional,

Table 10.1 Housing quality: The proportions of blocks at Bialla
and Hoskins with permanent, semi-permanent and bush material
houses

	Bialla (%)	Hoskins (%)
Permanent house	23	65
Semi-permanent house	63	78
Bush material house	82	78

Source: 2014 unpublished household survey data

political, socioeconomic and demographic factors. In the eyes of smallhold-ers, three key interrelated factors stand out: population growth on their leasehold blocks, competing demands on the oil palm income and the lim-ited opportunities to establish off-farm residences and livelihoods.

Population, land and economic pressures

At the schemes' inception, a 6-ha LSS block was deemed sufficient for the needs of a single family. However, the married couples who arrived on the schemes in the 1960s and 1970s with one or two young children now co-reside on their blocks with their married sons (and sometime daughters) and grandchildren. It is now common for three generations and several families to be sharing the resources of one 6-ha block. In 2014, the mean number of households per block at Bialla and Hoskins were 3.7 and 3.8, respectively. The growth of the multihousehold and multigenerational LSS block is reflected in the increased population density on the blocks. On the Hoskins LSS, population density has risen from 5.9 persons per block in the early 1970s (Ploeg 1972) to 8.6 in 2001 (Koczberski et al. 2001) and then to 18 persons per block in 2014 (and an average of 16 persons per block at Bialla in 2014).

Around the mid-1980s, the increasing population pressures on the LSS blocks came to the attention of provincial and national governments. In 1986, the East Sepik Welfare Association on the Hoskins LSS sent a delega-tion to the ESP to lobby the provincial government to seek national govern-ment support for an oil palm LSS in ESP. The aim was to address emerging land pressures on LSS blocks as sons married and raised their own fam-ilies on their parents' blocks (East Sepik Welfare Association 1986). In a 1992 field assessment report on land shortages on the LSSs in WNBP, the National Department of Agriculture and Livestock noted in their summary findings:

> The problem of over-crowding, land shortage, unemployment and increasing social problems does exist on the settlement blocks. All the concerned parties interviewed supported the moves by the settlers, not necessarily from East Sepik Province but from other province[s] as well, to be repatriated back to their respective provinces.
>
> Land shortage is indeed a problem and all concerned parties both at provincial and national level must act now to diffuse possible con-frontations between the traditional landowners and the settlers. The 6.5 ha block size is certainly not sufficient for settler family size of 10 to 13 children, inclusive of the second generation. The problem is indeed compounded by the current low prices for oil palm and existing land tenure covenants which restricts settlers to participate in other com-mercial activities.
>
> (Togiba 1992, p. 8)

Contributing to the rapid rate of population growth on the LSSs is the difficulty that settlers now experience when attempting to resettle in their "home" villages and the contraction of their prospects to establish residences and economic livelihoods beyond the LSS block. Opportunities for re-establishing themselves on customary land at "home" are becoming remote because of their long absences, together with the fact that many of their children were raised in WNBP and learned Melanesian Pidgin rather than their home/indigenous languages (see Curry and Koczberski 1999; Koczberski et al. 2001). Their home areas are also likely to be experiencing population pressure, given that settlers were initially recruited from land-short areas. Furthermore, the high rate of unemployment in PNG and the opposition to informal urban settlements by provincial governments (Koczberski et al. 2001) mean that settlers' off-block residence options are now much more constrained than in earlier decades. Whilst formal employment has increased in PNG since 2000, most has been in the mining sector. Urban employment opportunities, especially for youth, remain low (UNDP 2014). Thus many smallholders feel that there are few long-term sustainable solutions to population pressure on their blocks.

The increase in the number of co-resident households on the blocks also means that the fortnightly oil palm payments from the milling company are now spread across several co-resident households of varying age, status and household needs. Because each co-resident household must meet costs for healthcare, schooling, food and other basic necessities, these blocks often experience considerable economic and social pressures as per capita income from oil palm declines. It is now common for many households to rely on other sources of farm and non-farm incomes, such as selling vegetables at local markets, operating small retail enterprises or working as casual labourers (discussed further later in this chapter).

Declining per capita income from oil palm is a constant source of anxiety and tension on densely populated blocks. It often triggers arguments within and between household members, especially over the distribution of oil palm income on paydays. Arguments can erupt into violence between a father and his sons and between brothers. Disharmony amongst family members generates great anxiety, and smallholders may be preoccupied with solving family issues rather than being concerned with adopting good farm management practices. These disagreements are a major factor explaining low harvesting rates and low oil palm production amongst some families. Disputes over the remuneration of labour can disrupt harvesting and create disincentives to invest labour in good farm management practices.

There is also evidence that rising population on the LSS blocks is affecting educational attainment (Ryan 2015). There is a negative correlation between block population and education levels, suggesting that declining per capita incomes may be limiting educational opportunities (Table 10.2). Blocks with 31 people or more have on average 2.8 less years of schooling than people on the least densely populated blocks (1 to 10 people) (Table 10.2).

Table 10.2 Highest school grade achieved by block population for LSS smallholders (aged 15 to 49 years)

	Number of People on Block			
	1-10 (n = 28)	*11-20 (n = 227)*	*21-30 (n = 67)*	*31+ (n = 44)*
Average highest school grade completed	9.4	7.1	7.4	6.6

Source: Ryan 2015, p. 91

Economic and social stratification

A recent outcome associated with declining per capita incomes and competition over resources on the block is an erosion of cooperative, inter-household relations as co-resident households begin to act more autonomously. This is reflected in the increasingly common practice of denying some co-resident households regular access to oil palm income, leading to the emergence of economic and social stratification within co-resident households. Since 2010, our research indicates that co-resident households can be categorised as primary or secondary households. This division is based on a household's access to and control over the oil palm income, access to land for food gardening and their role in farm management decision-making. Primary households are those where the male head is typically the leaseholder and/ or has control over the distribution of the oil palm income and makes the major farm investment decisions on the block. Where the original leaseholder is deceased, typically the first-born son takes over as head of the primary household on the block. Secondary households are those headed by second- and subsequently-born married sons (and daughters) of the original leaseholder and/or relatives living permanently on the block (e.g., in-laws).

Secondary households' weak "ownership" claims to oil palm income are reflected in their income-earning activities. A larger proportion of secondary households than primary households are compelled to engage in non-farm income strategies. In 2014, approximately 32% of secondary households earned a regular income from wage employment compared with 19% of primary households. In addition, 45% of secondary households earned income from casual/informal farm labouring work compared with only 25% of primary households (Ryan 2015). Thus, members of secondary households with less access to oil palm income (and possibly land) are driven to pursue off-block employment.

Economic stratification is also evidenced by the higher education levels attained by children in primary households compared with secondary households who have less access to the monthly oil palm income (Ryan et al. 2013). The education levels of adults who receive monthly oil palm income from the Papa Card are higher than the lowest-ranked secondary

Table 10.3 Average years of schooling for second and third generation smallholders aged 15 to 49 years according to frequency of access to Papa and Mama Card income

Frequency	Papa Card only (years) (n)	Mama Card only (years) (n)	Papa and Mama Cards (years) (n)
Every month	7.3* (250)**	7.4 (208)	7.5 (172)
Never	5.6* (33)**	5.0 (17)	4.6 (14)

*$p < 0.05$ (see Ryan, 2015: 40 for the full test output).
**The independent samples t-test was performed only on the relationship between schooling and access to Papa Card income, as this was the only relationship where count numbers were above 20 for all categories.

Source: Ryan 2015

households who never receive Papa Card income. Table 10.3 shows a positive correlation between frequency of oil palm income and education levels (Ryan 2015). Smallholders who receive Papa Card income every month have 1.7 more years of education than those who never receive this income. For Mama Card income, the difference between those who receive oil palm income every month and never is 2.4 years.

Children in households with monthly access to both Mama and Papa Card income receive almost 3 years more education than children from secondary households that never receive oil palm income (Table 10.3). Papua New Guineans, with little or no formal education, are more likely to end up living in poverty than those with tertiary education (UNDP 2014). This relationship between income and education highlights the disadvantage and precariousness of households with restricted access to oil palm income.

Social change and modernity

In this environment of economic and social stress and increasing stratification, the long-established moral and social order is being challenged. For young married men experiencing these pressures, few have reason to uphold existing father-son hierarchal relationships in oil palm management that give fathers control of income distribution. Intergenerational cultural norms, especially a father's right to his son's unpaid or "underpaid" labour and control of the oil palm income, are being opposed increasingly by sons no longer content with the old ways. Inevitably, this has led to pressures to restructure social and power relations between fathers and sons, resulting in the decline of inter-household cooperative labour practices (Koczberski and Curry 2016).

The economic and demographic pressures on the LSSs have also acted to undermine the ability of many fathers to fulfil two key long-term cultural obligations to their sons: providing access to land and raising their bride-wealth. In the primarily patrilineal societies of PNG, a son can generally rely on his father to organise and negotiate access to land and arrange his bridewealth payment on the unstated condition he meets his filial obligations to his father. However, for several reasons, very few of the original leaseholders have been able to arrange access to land for their sons beyond the existing 6-ha block. Most leaseholders have lost access to land in their home villages because of their long absences (Koczberski et al. 2009). Also, most married sons were born in WNBP; many are not fluent in their home language and lack sufficient knowledge of their oral histories or genealogies to reintegrate successfully in their "home" villages and lay claim to the lands of their fathers.

Yet, there are few opportunities in WNBP to purchase a leasehold block, and their prices are beyond what most leaseholders can afford. Given the high and growing population pressure on the blocks and the financial expectations of, and obligations to, leaseholders' wives, children and extended family members, accumulating savings is very difficult. Thus, most original leaseholders can offer their sons only shared access to (and limited control of) the family's 6-ha oil palm block. In response, married sons on densely populated blocks are increasingly seeking to "purchase" nearby customary land to plant oil palm to secure their own family's future in WNBP. The deposit on the land "purchase" is often funded largely from their own savings, while remaining instalments are paid from the income they earn from the new oil palm stand when it comes into production. Thus, a central element in the social glue between fathers and sons, where a father secures access to land for his sons, is eroding with a consequent atomisation of household production.

Similarly, it appears that sons are relying less on their fathers to fund their bride prices. Many sons have partly funded their own bride-price payments through plantation work, oil palm income or small businesses on the block. Thus, as fathers have been increasingly unable to meet their paternal obligations, and sons partly fund and arrange their own bride-price payments and access to land, the social and moral order underpinning father-son relationships has been weakened with a shifting balance in the intergenerational relations of power and authority. The result has been a revaluing and a transformation of generational and social relations. It is here where the generational contexts and tensions on the LSS differ most from the past situation in PNG. Previously, young men/sons were dependent on older men for land and bridewealth, and this served to maintain the authority and dominant position of older men. This situation no longer pertains on the LSSs as more sons independently source land and bridewealth, thereby undermining both the seniority of their fathers and intergenerational power relations.

However, to attribute most present day uncertainties and pressures to the growing economic and population pressures on the leasehold block ignores other profound social changes that have occurred since the inception of these schemes throughout much of PNG. When settlers moved onto the schemes, their social lives and experiences were transformed as they began a new life where indigenous authority, social structures, clan identity and kinship networks were weak, where land titles were individualised and where commodity production was emphasised within ideologies of progress, individual autonomy and national development. In this environment, major markers and symbols of clan and ethnic identity have gradually lost their meaning. Among second- and third-generation settlers, Melanesian Pidgin has displaced indigenous languages, cash has replaced traditional items of exchange and everyday life has become increasingly disconnected from the cultural past of the first-generation settlers. Thus, these initial and ongoing social transformations, together with rapidly changing socioeconomic conditions occurring in most parts of PNG, appear to have contributed to people making choices that have expressed greater individualism and served to restructure household and generational social relations and identities.

Conclusion

This chapter has outlined the historical, socioeconomic and demographic contexts in which smallholders have made decisions regarding oil palm production and has examined the ways in which these settlers have responded to the opportunities, risks and emerging uncertainties on the ambitiously planned LSSs. On several socioeconomic indicators, the oil palm growing areas in WNBP rank highly relative to other rural areas of PNG. However, whilst the schemes have delivered monetary gains much greater than most settlers might have achieved in their home villages, the constraints of the leasehold block (fixed land area and restricted land use options) have created new challenges that were not imagined when settlers first moved onto their blocks. Processes of social change and modernity are ushering in new identities and reshaping social relations within and between households.

In the contemporary environment of the LSS, the optimistic view of the original settlers regarding the future possibilities of the schemes has been gradually replaced by a perception of the schemes as limiting their opportunities to improve their living standards. The original settlers who came to the LSS with high hopes and aspirations of living an imagined lifestyle of the Europeans were disappointed when that dream was not fully realised. However, social values and worldviews are changing for the new generation of settlers born and raised on the blocks. Most of them are reluctant to return "home", and many dislike the demands made on them by village relatives. They do not have the same spiritual and kinship attachments to "home" as do their parents, and they see their long-term futures in WNBP. They are more market-orientated in their relationships and often resist pressures from their families to value labour and social relationships from an indigenous economic perspective — they demand control over their own labour and market rates of return for their work in oil palm. Thus, the project of socioeconomic transformation initiated by the Australian administration and so eagerly embraced by the first generation of settlers may now, some 50 years on, be in progress amongst second- and third-generation settlers.

Notes

1. State agricultural leasehold land is state land alienated from customary ownership and leased by the state. Almost all state land was alienated prior to PNG's political independence in 1975. Alienated state land is removed from the regulation of custom and is "owned" according to the Land Act and other land laws of PNG (Larmour 1991).
2. PNG became self-governing on 1 December 1973, with full political independence on 16 September 1975.
3. The Melanesian pidgin term *bisnis* does not coincide exactly in meaning with the English word "business". *Bisnis*, while encompassing what is conventionally meant by "business", has a broader meaning and can include activities that have non-market elements such as *singsing bisnis*.
4. Yield averaged for three years (2014–2016) from data in PNGOPRA 2017:E7. Average price for fresh fruit bunches (FFB) was K239.27 in 2016.

References

Adamson, F, Fett, L, Huntsman, A and Scarlett, G 1984, *West Nakanai (Kimbe) Oil Palm Scheme, Papua New Guinea: Social, Economic and Environmental Aspects*, Environmental Report 20, Graduate School of Environmental Science, Monash University, Melbourne.

CILM (Commission of Inquiry into Land Matters) 1973, *Report of the Commission of Inquiry into Land Matters*, Government Printer, Port Moresby.

Connell, J 1997, *Papua New Guinea: The Struggle for Development*, Routledge, London.

Cramb, R and Curry, GN 2012, "Oil palm and rural livelihoods in the Asia-Pacific region: an overview", *Asia Pacific Viewpoint*, Vol. 53, No. 3, pp. 223–39.

Curry, GN and Koczberski, G 1999, "The risks and uncertainties of migration: an exploration of recent trends amongst the Wosera Abelam of Papua New Guinea", *Oceania*, Vol. 70, No. 2, pp. 130–45.

Curry, GN, Koczberski, G, Omuru, E, Duigu, J, Yala, C and Imbun, B 2007, *Social Assessment of the Smallholder Agriculture Development Project*, report prepared for the World Bank, Curtin University, Perth.

Densley, B 1980, "Rural policies: planning and programmes, 1945–1977", in D Denoon and C Snowden (Eds), *A Time to Plant and a Time to Uproot: A History of Agriculture in Papua New Guinea*, Institute of Papua New Guinea Studies, Port Moresby, pp. 285–93.

Diapo, R 1995, *A Brief Report on Bialla Smallholder Oil Palm Scheme and OPIC Extension Services*, unpublished Oil Palm Industry Corporation report, Baubata, West New Britain, Bialla.

East Sepik Welfare Association 1986, unpublished letter to the honourable premier and provincial members, East Sepik Provincial Government, 3 February 1986.

Fleming, T 1972, "The company view", in JP Longayroux, T Fleming, A Ploeg, RT Shand, WF Straatmans and W Jonas (Eds), *Hoskins Development: The Role of Oil Palm and Timber*, New Guinea Research Bulletin No. 49, Australian National University, Canberra, pp. 7–20.

Fold, N 2000, "Oiling the palms: restructuring of settlement schemes in Malaysia and the new international trade regulations", *World Development*, Vol. 28, No. 3, pp. 473–86.

Grieve, RB 1986, "The oil palm industry of Papua New Guinea", *Australian Geographer*, Vol. 17, No. 1, pp. 72–6.

Hulme, D 1983, "An economic appraisal of the Hoskins Oil Palm scheme", *Australian Geographer*, Vol. 15, No. 5, pp. 330–34.

Hulme, D 1984, *Land Settlement Schemes and Rural Development in Papua New Guinea*, unpublished PhD thesis, James Cook University, North Queensland.

IBRD (International Bank for Reconstruction and Development) 1965, *The Economic Development of the Territory of Papua and New Guinea*, John Hopkins Press, Baltimore.

Jonas, WJA 1972, "The Hoskins oil palm scheme", *The Australian Geographer*, Vol. 12, No. 1, pp. 57–8.

Kean, P 2000, "Economic development in the Siki settlement scheme, West New Britain", *Critique of Anthropology*, Vol. 20, No. 2, pp. 153–72.

Koczberski, G and Curry, GN 2003, *Sustaining Production and Livelihoods among Oil Palm Smallholders: A Socio-Economic study of the Bialla Smallholder Sector*, Research Unit for the Study of Societies in Change, Curtin University of Technology, Perth.

Koczberski, G and Curry, GN 2004, "Divided communities and contested landscapes: mobility, development and shifting identities in migrant destination sites in Papua New Guinea", *Asia Pacific Viewpoint*, Vol. 45, No. 3, pp. 357–71.

Koczberski, G and Curry, GN 2005, "Making a living: land pressures and changing livelihood strategies among oil palm settlers in Papua New Guinea", *Agricultural Systems*, Vol. 85, No. 3, pp. 324–39.

Koczberski, G and Curry, GN 2016, "Changing generational values and new masculinities amongst smallholder export cash crop producers in Papua New Guinea", *The Asia-Pacific Journal of Anthropology*, Vol. 17, Nos. 3–4, pp. 268–86.

Koczberski, G, Curry, GN and Gibson, K 2001, *Improving Productivity of the Smallholder Oil Palm Sector in Papua New Guinea: A Socio-Economic Study of the Hoskins and Popondetta Schemes*, Department of Human Geography, Research School of Pacific and Asian Studies, Australian National University, Canberra.

Koczberski G, Curry, GN and Imbun, B 2009, "Property rights for social inclusion: migrant strategies for securing land and livelihoods in Papua New Guinea", *Asia Pacific Viewpoint*, Vol. 50, No. 1, pp. 29–42.

Landell Mills Ltd. 1991, *Smallholder Oil Palm Productivity Study*, Department of Agriculture and Livestock, Papua New Guinea, Konedobu.

Larmour, P 1991, "Registration of customary land: 1952–1987", in P Larmour (Ed), *Customary Land Tenure: Registration and Decentralisation in Papua New Guinea*, Monograph 29, Papua New Guinea Institute of Applied Social and Economic Research, Port Moresby, pp. 51–72.

Longayroux, JP 1972, "Hoskins Oil Palm Project: an introduction", in JP Longayroux, T Fleming, A Ploeg, RT Shand, WF Straatmans and W Jonas (Eds), *Hoskins Development: The Role of Oil Palm and Timber*, New Guinea Research Bulletin No. 49, Australian National University, Canberra, pp. 1–6.

National Statistical Office of Papua New Guinea 1981, *National Population and Housing Census 1980*, National Statistical Office, Port Moresby.

Papua New Guinea Palm Oil Council 2012, *Palm Oil Industry Statistics*, Port Moresby.

Pincock, S 2006, "Papua New Guinea struggles to reverse health decline", *The Lancet*, Vol. 368, No. 9530, pp. 107–8.

Ploeg, A 1972, "Sociological aspects of Kapore settlement", in JP Longayroux, T Fleming, A Ploeg, RT Shand, WF Straatmans and W Jonas (Eds), *Hoskins Development: The Role of Oil Palm and Timber*, New Guinea Research Bulletin No. 49, Australian National University, Canberra, pp. 21–118.

PNGOPRA 2017, *PNG Oil Palm Research Association Annual Report 2016*, PNGOPRA, Dami Research Station, West New Britain Province, Papua New Guinea.

Post Courier 2007, "Kimbe bubbles with life", *Post Courier, Weekend Extra (Travel and Sports)*, 21–23 September, p. 4.

Reilly, B, Brown, M and Flower, S 2015, *Political Governance and Service Delivery in Papua New Guinea: A Strategic Review of Current and Alternative Governance Systems*, The National Research Institute, Discussion Paper No. 143, Papua New Guinea, Port Boroko.

Ryan, S 2015, *The Role of Status Hierarchies and Resource Allocation on Education Attainment of Papua New Guinea Oil Palm Smallholders*, MPhil thesis, Curtin University, Perth.

Ryan, S, Koczberski, G, Germis, E and Curry, GN 2013, *Smallholder Education Strategies Report. Developing a Smallholder Engagement Strategy for OPIC*, Curtin University and PNG Oil Palm Research Association, Perth.

Ryan, S, Koczberski G, Curry, GN and Germis, E 2017, "Intra-household constraints on educational attainment in rural households in Papua New Guinea", *Asia Pacific Viewpoint*, Vol. 58, No. 1, pp. 27–40.

SCLD (Select Committee on Land Development) 1971, *Report of the Select Committee on Land Development Together with Minutes of Proceedings*, National Parliament Library, Port Moresby.

Sutton, K 2001, "Agribusiness on a grand scale – FELDA's Sahabat Complex in East Malaysia", *Singapore Journal of Tropical Geography*, Vol. 22, No. 1, pp. 90–105.

Togiba, C 1992, *Resettlement of WNB Oil Palm Project Settler Back to ESP*, unpublished field visit report, Department of Agriculture and Livestock, Konedobu.

UNDP 2014, *National Human Development Report, Papua New Guinea: From Wealth to Wellbeing: Translating Resource Revenue into Sustainable Human Development*, United Nations Development Programme and PNG Department of National Planning and Monitoring, Port Moresby and Massey University, Auckland.

World Health Organisation 2015, "Country profile: Papua New Guinea," www.who.int/countries/png/en/ (accessed date: 20 June 2016).

11 The complexities of the Brazilian Amazon frontier and its most recent chapter: soybean expansion in Roraima

Alexandre M. A. Diniz
Elisangela G. Lacerda

Introduction

The state of Roraima lies in the northernmost part of Brazil, sharing borders with the Bolivarian Republic of Venezuela and the Republic of Guyana. It has direct road access to the Manaus metropolitan region in the state of Amazonas, a few hundred kilometers to the south. This strategic geopolitical situation presents considerable economic opportunities but also significant vulnerability to external influences, a factor that has played a paramount role in the history of the state (Figure 11.1).

Roraima has a short history of non-Indigenous occupation and economic exploration. It was not until the 1980s that a reliable terrestrial link with the rest of Brazil was achieved. Since then, Roraima has experienced tremendous population growth with thousands of migrants arriving in search of mineral riches, land in colonization projects or employment in the tertiary sector of Boa Vista, the capital city. Roraima still represents one of the last resorts for the Brazilian landless. It possesses active agriculture frontier areas, vast tracts of pristine lands, national parks and native peoples' reservations.

The BR-174 highway crosses Roraima in a north-south fashion, establishing a terrestrial link between Manaus and Paracaima, at the Venezuelan border; the BR-401 connects Boa Vista to Bonfim, at the border with Guyana (Figure 11.2). These roads have the potential to become important international corridors for goods and people. Nonetheless, given the current lack of economic vitality in these neighboring countries, this strategic position has been little explored, despite the formalization of joint international development and integration projects (Barros, Padula and Severo 2011). Despite its strategic location, Roraima has, to date, established only feeble and indirect links with global markets.

But this relative isolation is in the process of change. In response to the international appetite for commodities and mimicking processes witnessed in the southern fringes of the Amazon, Roraima experienced the progressive arrival of soybean plantations during the 1990s and 2000s as a result of

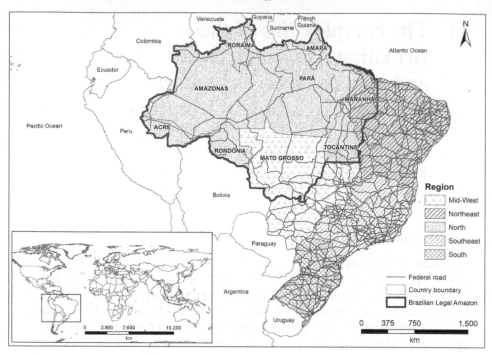

Figure 11.1 Brazil – the Amazonian context.

Source: The authors

its low land market values, favorable climate and its location. Roraima is much closer to major international consumer markets, and soybean can be easily shipped abroad through Venezuela and Guyana.

This chapter is based on secondary data and fieldwork that explore the recent historical and geographical expansion of soybean plantations in Roraima, examining some of their social, economic, cultural and environmental impacts. But, before we go any further, it is important to recognize that the spread of soybean in Roraima is an outcome of a series of wider developments associated with the Amazonian frontier and the evolution of Brazilian agribusiness, which we feel obliged to briefly discuss in order to provide some historical and geographical context.

The evolution of the Amazonian frontier

Frontier expansion in the Brazilian Amazon: the early years

The settlement history of the Brazilian Amazon is a relatively recent phenomenon, only dating back to the twentieth century. The region remained sparsely populated throughout the entire colonial period (1500–1822). The presence of natural resources with high world prices and demand was either

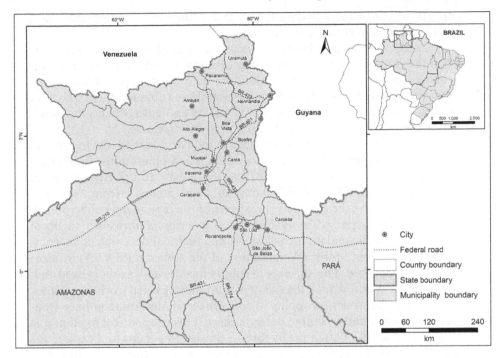

Figure 11.2 State of Roraima, regional context.

Source: The authors

unknown or unappreciated because the Portuguese were busy exploiting timber, developing sugar cane plantations along the coast and extracting minerals in south-central Brazil (Andrade 1989).

Nonetheless, the discovery of the vulcanization process, in 1845, changed the fate of the Brazilian Amazon substantially, sending hordes of migrants into the region to capitalize on the abundant sources of high-yielding, premium-quality latex rubber trees. Migrants originating in the semiarid and poverty-stricken Northeast area of Brazil flooded the region, establishing the Brazilian Amazon as the primary world supplier of rubber.

The incomers were predominantly young males, who used the tributaries of the Amazon River as penetration routes into the jungle in search of rubber trees. These unplanned and spontaneous population flows were intense, provoking a tenfold regional population growth – from an estimated 137,000 individuals in 1820 to 1.2 million inhabitants in 1910 (Schilling 1981; Oliveira 1983).

This unchecked migration led to a serious geopolitical issue when thousands of Brazilians were caught rubber-tapping deep in the Bolivian jungle. This Acre incident was diplomatically resolved by the Treaty of Petrópolis in 1903, by which this region was annexed by Brazil in exchange for the construction of the Madeira-Mamoré Railroad, which gave Bolivian rubber exports access to the Amazon River, and a financial compensation of 2 million pounds sterling (Andrade 1989).

The rubber boom lasted until the second decade of the twentieth century, when the Amazonian monopoly of this resource became increasingly compromised by Southeast Asian plantations. With the collapse of the rubber economy, massive out-migration took place; those rubber tappers who were reluctant to leave settled the local river plains and engaged in subsistence agriculture, gold mining and collection of forest products.

The fortunes of the Brazilian rubber industry turned during World War II, when Japanese forces took over the Asian plantations. Brazil then became the sole supplier for the Allies, leading to an upsurge in rubber businesses and an influx of 55,000 migrants, again mostly from Northeast Brazil. Inspired by the economic prospects of the revitalized rubber industry, President Vargas (1930–1945) created a series of measures to boost development in the region, including the first regional development agency, the Superintendence for the Valorization of the Amazon (SPVEA) (Mattos 1975; Andrade 1989). This upsurge in rubber businesses was short lived and, by the end of World War II, the Asian plantations regained world hegemony.

Since colonial times, geopolitical concerns over the Amazon have troubled Portuguese and Brazilian policymakers. This geopolitical fixation was shaped by a series of attempts by foreign powers to occupy and explore the region. Throughout the sixteenth and seventeenth centuries, for instance, the Brazilian Amazon was the focal point of secretive exploratory endeavors by the French, British and Dutch in search of exotic woods, oily seeds, fish and precious metals (Oliveira 1983). These forays prompted the official closing of the Amazon River by the Portuguese in 1616.

A much more recent foreign response to the geopolitical obsession with the Amazon was UNESCO's attempt to create the so-called International Institute for the Amazon in 1948, which was designed to direct and support scientific research in the region. During the 1960s, the region experienced another blow to its territorial sovereignty, when the Hudson Institute called for the internationalization of the Brazilian Amazon as a means of increasing South American development (Reis 1968; Mattos 1980).

This historical background, coupled with the region's vast scale, isolation and exceptional and unknown riches inspired a series of strategies to keep the Amazon under Brazilian control. President Kubitcheck (1956–1961) led the most audacious geopolitical intervention ever witnessed in Brazil by transferring the national capital from Rio de Janeiro to the recently built Brasília in 1961. Among other things, the new capital city was expected to induce the economic and political integration of the country by transferring population from densely inhabited coastal areas to the empty interior. President Kubitcheck also held on to SPVEA's propositions, promoting the construction of Belém-Brasília road (BR 153), the very first terrestrial link between the Amazon region and South-Central Brazil (Andrade 1989; Costa 1991) (Figure 11.1). This road proved to be instrumental in the demographic occupation and the economic development that took place in the following decades, including the expansion of soybean plantations.

The era of centralized and inconsistent regional plans:
the military administrations of 1964–1985

The military coup of 1964 brought to power an autocratic and national-ist regime whose plan of action was inspired by the doctrine of national security elaborated within the Escola Superior da Guerra (Superior War School) in the late 1940s. The military government embraced positivism and envisioned the Brazilian future as an affluent, scientific, rational society marked by heavy social planning and beneficial authoritarian governments (Foresta 1992). The two pillars of the doctrine – economic development and national security – fueled various centrally planned development meas-ures within the next 20 years, which heavily impacted the Amazon region (Mattos 1975).

Under General Castello Branco (1964–1967), the first post-coup president, "Operation Amazônia" set the tone for regional development. Its backbone was Law 5.1744 (1966), which granted fiscal incentives to entrepreneurs by stipulating that 50% of corporate tax liability could be invested in devel-opment projects in the Amazon region. This new law essentially turned taxes into venture capital. Corporate entities benefited at the time because they had the administrative and financial capability to address the massive regional infrastructural problems that otherwise would have to have been overseen by the government.

As part of Operation Amazon, the Superintendency of Development for the Amazon (SUDAM) replaced the old SPVEA, becoming the chief organization responsible for the economic development of the region (IBGE 1991). The new agency was also charged with the responsibility of assisting the private sector in the construction and improvement of transportation routes. The paving of the Belém-Brasília road (BR 153) and the opening of a terrestrial link between Cuiabá and Porto Velho (BR 364) epitomized some of its accomplishments (Mahar 1979) (Figure 11.1).

Fiscal incentives and abundant credit lines resulted in a plethora of national and transnational ventures in the Amazon, the bulk of which tar-geted cattle ranches in the northern Mato Grosso and southern Pará states, areas closer to major markets and with good transportation infrastructure (Figure 11.1) (Sawyer 1981; Hecht 1984; Correa 1987; Valverde 1990).

By the late 1960s, criticism over the expansion of corporate livestock operations mounted as regional agriculture, which had been historically based on small-scale extractive industries and subsistence agriculture, was transformed into a profit-driven *latifúndio*[1] system, reproducing the land structure of the more settled Brazilian regions. Concomitantly, post-war industrialization was abrupt and promoted rapid modernization of urban and rural areas, generating technological unemployment and displacing thousands of sharecroppers and rural workers. Large-scale production also dampened the prices of agricultural goods, running many small-scale producers out of business (Graziano 1996). Nevertheless, the official response to the centuries-old land concentration problem and to the side effects of fast agricultural modernization was not serious agrarian reform, as claimed by many, but the further colonization of the Amazon region.

Within this context, President Médici (1969–1974) launched PIN I (First National Integration Plan), which called for the construction of the Transamazônica (BR 364) and Cuiabá-Santarém (BR 230) roads and the creation of a series of colonization settlements along these transportation routes. The primary beneficiaries of this initiative were to be the landless peasants[2] of Northeast Brazil (Mattos 1980; Hecht 1984; Henriques 1984) and a newly founded agency, the Instituto Nacional de Colonização e Reforma Agrária (INCRA – National Institute for Colonization and Agrarian Reform), which was charged with the responsibility of settling the families. INCRA envisioned the settlement of 100,000 families between 1971 and 1974, with financial support from SUDAM.

Nonetheless, PIN I was a complete failure as, by 1974, only 5,717 families were settled (Arruda 1978; Sawyer 1981), of which only 40% still held on to their plots in 1980 (Henriques 1984; Correa 1987; Fearnside 1989). Among the reasons for this fiasco were the peasants' low levels of education; lack of training and experience with agriculture in the Amazon region; low agricultural productivity; inadequate provision of seeds and other inputs; lack of reliable transportation linkages and storage facilities; delays in the provision of land titles, which hindered peasant access to credit and a land demarcation process that overlooked hilly topography, access to water and soil quality (Henriques 1984; Stone 1985; Foresta 1992).

Loud criticism and fierce opposition against PIN I, coupled with OPEC's oil embargo, forced the Brazilian government to divert resources from the projects underway in the Amazon toward the purchase of oil, leading to a severe rupture in regional planning (Mattos 1980). Policy focused on the occupation and development of the region by peasants and small-scale farmers was changed in favor of encouraging corporations and capitalized farmers, players who had the means to promote faster and cheaper occupation of the region. Tax breaks and funds were made available at very low interest rates to stimulate new projects (Fearnside 1989; Becker 1990; Valverde 1990) while INCRA auctioned large lots of land (Pinto 1997), which were acquired under highly advantageous conditions (Sawyer 1981; Henriques 1984; Becker 1989). This policy change led to the further expansion of cattle ranches, which represented the bulk of the agricultural projects developed during this period.

The late 1970s were marked by soaring interest rates worldwide, which raised the Brazilian external debt to more than US$100 billion. The need to meet these external payments inspired a different strategy for the Amazon region, whose mineral riches were about to be exploited in an unprecedented fashion. President Geisel (1974–1979) launched the Polamazônia program (Amazonia's agricultural and mineral development poles program), which was designed to catalyze agro-ranching and mineral development poles throughout the region and was based on consortia between the Federal government and the private sector. As a result, 1,887 mineral or agro-ranching projects were approved by SUDAM between 1975 and 1986 (IBGE 1991).

President Geisel's emphasis on natural resource exploration was further advanced by President Figueiredo (1979–1985). Prospecting for minerals and developing these resources became the trademark of the Amazon during his mandate, including the Carajás (iron ore and manganese) and Trombetas (bauxite) mines, the Tucuruí Dam (hydroelectricity), the Albrás plant (aluminum), the Alunorte refinery (alumina) and the Alunar smelter (aluminum) (Becker 1989). The implementation of these projects drew many construction and industrial workers from different parts of Brazil, adding some demographic complexity to the regional migrant stocks.

One must not forget the ubiquitous presence of small-scale, independent and, for the most part, clandestine *garimpeiros*[3]. Following the course of rivers, these miners surveyed large tracts of land in search of gold and diamond veins, which were exploited in subhuman conditions with a great deal of environmental degradation. The number of *garimpeiros* was estimated at 10,000 in 1960, contrasting with 240,000 by 1990. This dramatic increase was also an outcome of the population pressure over local resources, as *garimpos*[4] became an economic alternative for late-arriving peasants who could not get access to land and for urban workers without stable employment (Godfrey 1992).

Geopolitics and the plight of landless Brazilians were not neglected by President Figueiredo who, in the mid-1980s, launched the Polonoroeste (Northwest Region Integrated Development Program), targeting the demographic occupation of the northwestern fringes of the Amazon. Funded by the World Bank, this project sought to rebuild and pave the Cuiabá-Porto Velho road (BR 364), improve existing colonization projects and create new ones (IBGE 1991). Polonoroeste had a tremendous demographic impact in the state of Rondônia, especially after construction of the roads revealed patches of premium soils, which attracted thousands of migrants not only from Northeast Brazil but also from the south-central portions of the country (Figure 11.1).

Neoliberalism and globalization in the frontier: the democratic era

In 1986, the so-called "New Republic" inaugurated a new democratic order under civil administrations and with regular elections. President Sarney (1986–1990) took power promising to curb the gigantic social and environmental problems plaguing the Amazon region at the time. His response was the I PDR-NR (First Regional Development Plan for the Amazon), which sought to catalyze regional economic growth while maintaining ecological balance and reducing social inequalities. Despite the strength of its political rhetoric, the plan was short-lived given the lobbying of conflicting interest groups.

Environmentalists were among these lobbying parties, and they possessed increasing power given the international outcry against the widespread burning and vast environmental degradation generated by the ongoing frontier expansion. In response to the domestic and foreign commotion in the region provoked by traditional agriculture and ranching activities, President Sarney unleashed various environmentally related initiatives. In 1988, an agency to enforce environmental laws (IBAMA – Environmental Brazilian Institute) was created along with the *Nossa Natureza* (Our Nature) plan, with the goal of restricting the "predatory exploitation of Amazonian resources". Notwithstanding its positive reception by environmentalists and the international community, the program did not work as anticipated.

Geopolitical concerns persisted as the armed forces remained influential within the civilian government, dominating the National Security Council. Motivated by the need to protect Brazil from the domestic disorder occurring in neighboring countries, the *Calha Norte* project was launched in 1985, affecting some 6,500 km[2] along the international boundaries with Colombia, Venezuela, Guyana, Suriname and French Guyana. The plan established military colonies, improved transportation in the northernmost territories and introduced economic growth poles based on large-scale mining activities at the same time as intensifying diplomatic relations with these neighboring countries.

Also noteworthy during this period was the discovery of gold deposits in the Yanomami territory in 1987, which prompted the arrival of 40,000 *garimpeiros* in the state of Roraima (Figure 11.1). The government tolerated the intrusion because they saw the *garimpeiro* invasion as a good opportunity to "civilize" the Indigenous population and curb the geopolitical problem that they were perceived as representing[5].

President Collor (1990–1992) took office in 1990, inheriting the Yanomami controversy. Pressured by public opinion, Collor took more aggressive action, expelling the *garimpeiros* in an unprecedented operation that was applauded by environmentalists and human rights advocates. Collor also attempted to neutralize the most salient and anachronistic aspects of the military hegemony in the Amazon by becoming more active in the defense of native population groups and the environment (Albert 1992). With Collor's impeachment, Vice President Franco (1992–1994) took over the presidency in 1992, but had little time to implement any large-scale plans for the region.

President Cardoso (1995–2002) established yet another set of contradictory measures related to the Amazon as he attempted to accommodate the competing ideals of regional development, global market integration and environmental conservation. During his first year in power, a special secretariat dedicated to the affairs of the Amazon region was created within the Ministry of the Environment with the mission of reformulating the environmental policies for the Amazon region in order to balance conservation and sustainable forms of economic exploitation. Although generating promising results, these initiatives were restricted to pilot projects of limited scope (Serra and Fernández 2004).

Moved by the historical contexts of globalization, liberal reforms and the organization of international trading blocs, Cardoso reenacted broad, centrally planned actions with the *Brasil em Ação* (1996–1999) and *Avança Brasil* (2000–2003) plans. These sought to modernize productive structures and eliminate bottlenecks in the productive and commercial chains and to increase the competitiveness of national products in global markets (Oliveira 2015).

These plans closely resembled the interventions devised by the military governments of the 1960s and 1970s. Nonetheless, instead of working with the poles of development approach, Cardoso's plans adopted an axial logic of development, placing heavy emphasis on transport infrastructure to improve internal regional articulation. Another important difference from previous territorial policies was the attention paid to integration with the rest of South America as well as to opening the markets of the Atlantic to those of the Pacific. Among the nine designed integration axes, three were in the Amazon region (Arco-Norte, Madeira-Amazonas and Araguaia-Tocantins), where the improvement of terrestrial communication was intended to facilitate the transport of locally produced commodities such as timber, minerals, beef and soybean, improving production and transportation costs and increasing domestic and international competitiveness (Superti 2012).

By the mid-1990s, the Amazon was plagued by a panoply of problems, including drug trafficking, smuggling, unorganized peasant colonization, invasion of Indigenous lands, illegal mining, timber exploitation and widespread ecological devastation. The enormous size of the region, coupled with difficulties in communication, made knowledge of and control over legal and illegal activities virtually impossible. Seeking to regain control and understanding of the processes occurring in the Amazon, President Cardoso launched the SIPAM/SIVAM[6] project, with clear defense intentions. However, the neoliberal waves that swept through Brasília had diminished the effectiveness of the federal government in regional development affairs. The "minimum state" philosophy, coupled with the dubious manner in which Raytheon won the bidding process to supply equipment for the surveillance of the Amazon, have slowed the implementation of this program (Contreiras 1998).

With the ever-increasing budget cuts promoted during President Cardoso's rule, SUDAM lost most of its power and ability to significantly change the course of activities in the Amazon. As a matter of fact, many claim that Brasília has turned its back on the Amazon. This more laissez-faire approach toward the region has led to even more-pronounced environmental degradation. The Instituto Nacional de Pesquisas Espaciais (National

Institute for Space Research) estimates that deforestation in the Amazon region amounted to 78 thousand square kilometers between 1995 and 1998 (INPE, 2018).

With the arrival of President Lula (2003–2010), a development strategy was adopted in Brazil that combined economic, social and cultural promotion, especially of the poorest (Pochmann 2011; Ribeiro 2016). In contrast to previous decades, which were marked by low economic growth, hyperinflation and economic instability, the 2000s witnessed consistent economic growth, catalyzed mainly by the appreciation of commodity prices in the international market and the growing government budget surplus (Maia 2013). This neodevelopmentalist approach has produced significant advances in economic and social indicators, especially decreasing inflation rates, an increase in the number of formal jobs, a significant increase in the mass of labor income and the reduction of social inequalities. In addition to economic growth and the generation of jobs, the introduction of and adherence to a minimum wage policy has significantly increased workers' purchasing power, and the creation or expansion of income transfer policies has benefited historically excluded segments of the population (Ribeiro 2013).

Like his predecessor, President Lula also emphasized international integration strategies in the large PPA (2004–2007) and Growth Acceleration Program (PAC, 2007–2010), which were designed to boost energy, transportation and communication infrastructure (Superti 2012). In the Amazon, the paving of several highways such as BR 163 (Cuiabá-Santarém) and the BR 319 (Manaus-Porto Velho) aims to facilitate the outflow of agricultural production and to make agricultural production more economically advantageous. This has the potential to increase both deforestation and conflicts over land tenure in areas now covered by the forest (Figure 11.1). These interventions have the potential to completely transform the region, altering the logistics and, consequently, the transport costs, especially those related to soybean – a cash crop that has already expanded throughout the southern fringes of the Amazon and is now encroaching to the north (Oliveira 2015).

The main investments of the PAC in the area of hydropower are at Belo Monte, on the Rio Xingu, and Santo Antônio and Jirau, on the Madeira River. Together, these three plants should generate more than 11,000 MW. In the area of energy transmission, 4.721 km of transmission lines were anticipated, connecting the energy generated in the Madeira River to the state of São Paulo in Southeast Brazil. Also with regard to energy infrastructure, two sections of the gas pipeline that will transport gas between Porto Velho and Manaus are being implemented in the state of Amazonas (Urucu-Coari and Coari-Manaus) in addition to a stretch that will connect Porto Velho (RO) to Ururucu (AM), which is still in the study phase (Quintslr et al. 2011).

Parallel to these broad infrastructural interventions, the 2000s also witnessed an expansion in the number of colonization projects in the Amazon. As the land concentration problem remained unsolved, successive federal administrations turned to the readily available lands of the Amazonian frontier to relieve the social problems besetting the backward rural areas elsewhere in Brazil. INCRA was called into action once again and created 794 agricultural colonies in the Amazon, benefitting 187,510 families during President Cardoso's term. President Lula was even more active in promoting agricultural settlements in the region, benefiting 293,986 families across 967 colonies (Mattei 2013).

Amazonian soybean

As one can grasp from this brief historical overview, the Amazonian frontier is a complex mosaic of multiple forms of land use and occupation. It is an open battlefield, where national and transnational corporate capitalists, hunter and gather Indigenous groups, ranchers, mining companies, peasants, rubber tappers, *garimpeiros*, logging companies, environmentalists, religious groups, alternative esoteric societies and so on fight over the control of local resources. But, among these players, one group has been particularly active in dominating regional affairs: the powerful agribusiness operations engaged in the production of grains and, most especially, soybean.

But, before we turn to the specifics of soybean plantations in the Amazonian frontier, we must build a broader historical context. Brazil is the second-largest producer of soybeans in the world, behind only the United States. The 2016 to 2017 harvest yielded 114,075 million tons (Table 11.1) (CONAB 2017), an amount achieved mostly by growing international demand. Soybean's wide use in industry, with applications ranging from food production for both human and animal consumption to the production of cosmetics and biodiesel, have made this the fourth most-consumed and produced agricultural product and a major oil crop (Hirakuri and Lazzarotto 2014).

Soybeans are believed to have been introduced to Brazil in the nineteenth century by Japanese immigrants who settled in the state of São Paulo. However, soybean production only became significant after its introduction into southern Brazil (Rio Grande do Sul and Santa Catarina) during the 1940s and 1950s (Brown et al. 2005). A major turn of events took place in the 1970s when a breakdown in the Peruvian production of anchovies prompted its replacement by soybeans as a major source of protein for animal feed, especially in North America and Europe (Fearnside 2001). This situation, coupled with a favorable international environment, a reduction in grain harvests in Russia and the inability of the United States (the world's largest producer of soybeans) to meet international demands, greatly stimulated the expansion of the crop in Brazil (Missão 2006).

During the 1970s and 1980s, soybean plantations spread to Central-West Brazil, taking advantage of the strategic characteristics of the Cerrado biome, a tropical savanna ecoregion with flat relief that was suited to mechanized farming; favorable climatic conditions with plentiful rainfall and

Table 11.1 Brazil – soybean production (in thousands of tons), 1990 to 2017

Region/Federal Unit	2010/11	2011/12	2012/13	2013/14	2014/15	2015/16	2016/17
NORTH	**1.977,20**	**2.172,20**	**2.661,50**	**3.391,30**	**4.289,50**	**3.818,90**	**5.536,40**
Roraima	10,4	10,4	33,6	56,2	63,9	79,2	90
Rondônia	425,3	462,2	539,3	607,7	732,9	765	930,3
Acre	-	-	-	-	-	-	-
Amazonas	-	-	-	-	-	-	-
Amapá	-	-	-	-	-	-	54,4
Pará	314,4	316,7	552,2	668,6	1.017,00	1.288,00	1.635,30
Tocantins	1.227,10	1.382,90	1.536,40	2.058,80	2.475,70	1.686,70	2.826,40
NORTHEAST	**6.251,50**	**6.096,30**	**5.294,80**	**6.620,90**	**8.084,10**	**5.107,10**	**9.644,70**
Maranhão	1.599,70	1.650,60	1.685,90	1.823,70	2.069,60	1.250,20	2.473,30
Piauí	1.144,30	1.263,10	916,9	1.489,20	1.833,80	645,8	2.048,10
Ceará	-	-	-	-	-	-	-
Rio Grande do Norte	-	-	-	-	-	-	-
Paraíba	-	-	-	-	-	-	-
Pernambuco	-	-	-	-	-	-	-
Alagoas	-	-	-	-	-	-	-
Sergipe	-	-	-	-	-	-	-
Bahia	3.507,50	3.182,60	2.692,00	3.308,00	4.180,70	3.211,10	5.123,30
CENTRAL-WEST	**33.938,90**	**34.904,80**	**38.091,40**	**41.800,50**	**43.968,60**	**43.752,60**	**50.149,90**
Mato Grosso	20.412,20	21.849,00	23.532,80	26.441,60	28.018,60	26.030,70	30.513,50
Mato Grosso do Sul	5.169,40	4.628,30	5.809,00	6.148,00	7.177,60	7.241,40	8.575,80
Goiás	8.181,60	8.251,50	8.562,90	8.994,90	8.625,10	10.249,50	10.819,10
Distrito Federal	175,7	176	186,7	216	147,3	231	241,5
SOUTHEAST	**4.622,10**	**4.656,30**	**5.425,90**	**5.015,30**	**5.873,50**	**7.574,90**	**8.151,50**
Minas Gerais	2.913,60	3.058,70	3.374,80	3.327,00	3.507,00	4.731,10	5.067,20
Espírito Santo	-	-	-	-	-	-	-
Rio de Janeiro	-	-	-	-	-	-	-
São Paulo	1.708,50	1.597,60	2.051,10	1.688,30	2.366,50	2.843,80	3.084,30
SOUTH	**28.534,60**	**18.553,80**	**30.025,80**	**29.292,80**	**34.012,30**	**35.181,10**	**40.592,80**
Paraná	15.424,10	10.941,90	15.912,40	14.780,70	17.210,50	16.844,50	19.586,30
Santa Catarina	1.489,20	1.084,90	1.578,50	1.644,40	1.920,30	2.135,20	2.292,60
Rio Grande do Sul	11.621,30	6.526,60	12.534,90	12.867,70	14.881,50	16.201,40	18.713,90
BRAZIL	**75.324,30**	**66.383,00**	**81.499,40**	**86.120,80**	**96.228,00**	**95.434,60**	**114.075,30**

Source: Companhia Nacional de Abastecimento (CONAB), 2017

sunlight; a natural vegetation composed of less bulky plant species, whose removal was far less costly than that for forested areas and wide availability of land at low prices (Bertrand, Cadier and Gasquès 2005). But the Brazilian state also gave an important push by funding a research agency Empresa Brasileira de Pesquisa Agropecuária (EMBRAPA), which became the chief body responsible for the genetic improvement and adaptation of soybeans to the environmental conditions present in the Cerrado biome, and for the heavy investments in transport infrastructure and the creation of tax breaks and credit lines to boost agricultural production. Further important support came from corporate agents themselves, who provided technical know-how and the financial capital necessary, and from the emergence of a dynamic and efficient cooperative system, which strongly supported the production, the industrialization and the commercialization of the crops (EMBRAPA 2004).

By 1990, soybean plantations were well established in central Brazil, from where they slowly began to encroach into Northeast Brazil and the southern fringes of the Amazon, most noticeably in the state of Mato Grosso, which, by that date, had become the third-largest producer in Brazil, with 1.6 million hectares planted (Domingues and Bermann 2012; Melo et al. 2015). In the following decades, soybean diffusion continued across the Amazon region as plantations slowly reached Rondônia, southern Pará, and, more recently, the northernmost Brazilian states: Amapá and Roraima (Figure 11.3).

Although the production of grains (initially maize and rice) had been an integral part of Amazonian frontier activity since the first large-scale colonization efforts by the military governments in the 1970s, over recent decades, the changing scale on which these commodities have been produced is noteworthy, especially when one takes into consideration the ever-expanding soybean plantations. Within just a few decades (1990–2010), the area devoted to soybeans in the Brazilian Amazon went from 1,573,404 ha to 6,995,455 ha, representing a 345% growth. Concomitantly, rice saw its lands shrink from 1,316,938 ha to 956.589 ha, while maize grew 157%, going from 1,055,339 ha to 2,715,001 ha. In addition to this tremendous growth in area cultivated, agricultural output has increased accordingly. In 1990, the Brazilian Amazon produced 5,799,580 tons of rice, maize and soybeans. By 2010, these grain outputs had reached 32,602,716 tons, representing a 462% growth. Nonetheless, of the total amount of grains produced in the Brazilian Amazon, 64.16% are soybeans, and Mato Grosso state, in the southern fringes of the region, is the chief producer (Silva 2015).

The massive investments in transport infrastructure developed during the Cardoso and Lula eras mobilized national and multinational agribusiness corporations who attempted to secure benefits and explore opportunities by either relocating, expanding or diversifying the production of grains in the Amazon, which, in turn, led to a redesign of production chains and export

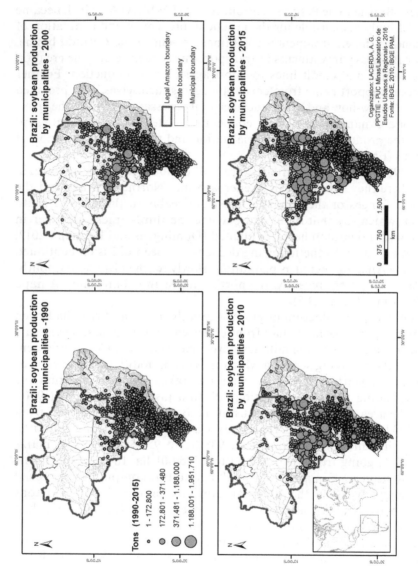

Figure 11.3 The evolution of soybean production in Brazil.

Source: The authors

channels, generating a new corporate geography. Oliveira (2015) shows how these corporations have carved up the Amazonian space:

- Rondônia state: Cargill and Amaggi have come to dominate production markets, by securing the use of the Rio Madeira waterway.
- Amazonas and Roraima states: embryonic grain production falls under control of Amaggi, which operates an industrial plant and a river harbor in Itacoatiara-Amazonas.
- Pará state: Cargill commands production in Southwest Pará and the lower Amazon, based on their port terminal in Santarém; while Bunge dominates east and Southeast Pará, based on their processing plants in Paragominas and Santana do Araguaia.
- Maranhão: Bunge is present in the northern and western parts.
- Tocantins: Bunge dominates, based on its units in Pedro Afonso, Guarai, Porto Nacional and at Gurupi e Campos Lindos, where Cargill is also present.
- Mato Grosso: as almost 90% of the Amazonian soybeans are produced in this state, all major corporations (Cargill, Amaggi, Bunge and ADM) are present here and are engaged in fierce disputes.

The arrival of soybeans brought about profound transformations in local production systems and social relations. Traditionally, the areas now under the dominance of the soybean industry were marked by isolation, subsistence or low surplus multi-crop agricultural systems, where extensive activities and low technology levels were applied to relatively small land parcels. Within a matter of years, these areas were appropriated by corporations supported by governments and regional elites, which adopted monocultural production processes based on large tracts of land; the adoption of technological innovations – be they mechanical, physical-chemical or biological; high standards of competitiveness and profitability and good connections to national and transnational markets, thereby virtually transforming the region into corporate territory (Soler, Verburg and Alves 2014; Silva 2015).

As we attempt to connect the settlement history of the Brazilian Amazon with these latest developments, it is possible to discern a clear succession of economic uses with increasing impacts on local ecosystems. The frontier exploration cycle initiates the extraction of natural resources, in a legal or irregular fashion, with an original emphasis on the extraction of timber. Once the area has been deforested, artificial grasslands are introduced along with cattle ranches. Depending on the location and on the evolution of the transport infrastructure, monocultures, such as soybeans, are introduced and their diffusion through space happens at a fast pace. In the final part of this chapter, we concentrate on this latter process.

Roraima represents the most recent chapter in the expansion of the Amazonian frontier. A region of late colonization that, beginning in the 1970s, became a focal point of intense federal government policies such as the construction of important federal highways, including BR-174, and the creation of rural settlements seeking to protect international border areas and stimulate economic development. Despite many plans and initiatives, Roraima's economy did not take off until very recently, and it remained heavily dependent upon federal public funds and the local service sector, which is, for the most part, restricted to public administration activities at the *município* and state levels. The primary sector is based on the embryonic extraction of forest products, namely timber. The agricultural sector is heavily dependent on subsistence farming systems, except for a sizable rice production operation near the northern border, which is currently immersed in serious land conflicts that were provoked by the demarcation of the Macuxi Indigenous reserve – Raposa Serra do Sol – in 2005.

But the prelude to important structural changes is already noticeable. Since 2010, soybeans have increasingly become an important source of revenue in Roraima. Producers from Central-West Brazil who are fleeing increasing land costs have noted Roraima as an opportunistic location in which to expand their businesses. The outlook is that production will increase further in the coming years, given rising world demand and Roraima's geographical location, which enjoys direct access to Venezuela and Guyana and, therefore, close proximity to Caribbean ports (Vera-Diaz, Kaufmann and Nepstad 2009). We now turn to the examination of soybean expansion in Roraima, but first we present some general aspects of the state that some like to call the last Brazilian frontier (Diniz 2002; Diniz 2003; Santos and Diniz 2004).

Roraima State

Roraima is one of the nine political units falling within the Brazilian Amazon river basin. Its landscapes encompass flat areas, mountains, forests and savannas. The savanna ecosystem, known locally as *lavrado*, covers approximately 17% of Roraima, spreading from the center to the northeastern fringes of the state (Gianluppi and Waquil 2008). Despite its large size (224.299 km²), relatively few areas are presently available for agricultural development because much of Roraima is under demarcated Indigenous lands or conservation units.

Roraima mirrors many current and past features and problems of other Amazonian areas in that it has experienced rapid development, massive road building, colonization programs, competition for land, destruction of natural vegetation, conflicts between Indigenous groups and settlers and an intense urbanization process (Furley and Mougeot 1994). Given its remote location, Roraima remained isolated from the rest of Brazil for centuries until the Portuguese implemented the first attempts to establish strongholds there in the eighteenth century.

Roraima's isolation was interrupted during the rubber boom, when the area became the chief source of beef produced in the cattle ranches located in the savanna region (Souza 1986; Silveira and Gatti 1988; Barros 1995; Freitas 1997). This temporary economic upsurge brought numerous migrants from poverty-stricken, semiarid Northeast Brazil seeking employment in the thriving beef industry. Notwithstanding this economic bonanza, Roraima was able to retain few people because cattle ranching was not a labor-intensive industry. Estimates suggest that Roraima had around 10,000 inhabitants by 1900. With the demise of the rubber economy, many left and, by 1920, only 7,424 individuals remained (Silveira and Gatti 1988).

During the 1930s, mining became the chief economic activity with the discovery of gold and diamonds along the River Cotingo. At the same time, diamond mines were discovered in the Tepequém mountains (Vieira 1971). These discoveries revitalized the economy and raised the population figure to 10,509 by 1940 (Silveira and Gatti 1988; Freitas 1997).

President Vargas implemented a series of measures to boost economic development in the Amazon region, the most relevant for Roraima being the implementation of the first agricultural colonization projects, which promoted the transference of hundreds of peasants from Northeast Brazil. Nonetheless, these colonization attempts were frustrated by tropical diseases, ineffective planning and the general lack of infrastructure, which discouraged permanent settlement. Still these processes further added to the local population, which reached 18,116 by 1950 and 28,304 by 1960 (Magalhães 1986; Souza 1986; Silveira and Gatti 1988).

Despite these developments, Roraima remained sparsely populated and economically isolated. A major impediment for the occupation and development of the territory was its sole reliance for transportation upon the Rio Branco upon which navigation during the dry season (October to April) was challenging if not impossible. This shortcoming was resolved in 1976 when a road (BR 174) linking Boa Vista, the state capital, with Manaus established the first terrestrial transport link with the rest of Brazil. The road was eventually extended to the Venezuelan border in 1998. It is also worth mentioning the construction of the Perimetral Norte (Northern Perimeter) road, which opened up the isolated lands of southern Roraima, as part of the geopolitical maneuvers led by the military governments discussed in the previous section (Figure 11.2).

These roads mark the beginning of a new era in the occupation of the region. Besides fostering a year-round connection with the rest of the country, they opened up vast tracts of "new lands" that had been explored in various colonization projects. As a result, the population, which was slightly over 28,000 inhabitants in 1960, reached 40,885 in 1970. This upward trend continued during the 1970s and, by 1980, the population reached 79,159.

The discovery of gold and diamonds in the northern portions of Roraima in the mid-1980s drew thousands of *garimpeiros* to the area. It is estimated that more than 40,000 individuals were directly involved in the *garimpo* rush between 1987 and 1991, not to mention those involved in the maintenance of the mining activity such as cooks, store clerks, pilots and so forth (Macmillan 1995). Owing to the intense mining activity, Roraima's population grew a staggering 10.64% a year during the 1980s, nearly tripling its population. Roraima was the fastest-growing Brazilian state at that time, reaching a population of 217,583 by 1991 (IBGE 1992).

The mining activity was mostly conducted in a clandestine fashion in Indigenous lands and, under pressure from foreign governments and environmental and human rights agencies, the federal government halted illegal mining in 1990. The closing of the *garimpos* provoked a significant migratory reflux, dramatically slowing the population growth. Yet, spontaneous migratory movements continued and, by the year 2000, the local population had reached 324,397 inhabitants, further advancing to 450,479 in 2010.

Despite significant population growth over recent decades, Roraima's economy remains underdeveloped and is still heavily reliant on federal public funds and administrative sector jobs. Until the 2000s, its exports were centered on timber, and despite numerous efforts to establish colonization settlements, agricultural production was meager. But that is in the process of changing with the arrival of soybean plantations, which we discuss in the next section.

Roraima's soybean frontier

Roraima soybean production is a relatively recent phenomenon. The first attempts to introduce the grain into the state date back to 1993. However, the low productivity of plant varieties, coupled with limited access to agricultural inputs, caused production to be abandoned. During the 2000s, new attempts were made, but yields oscillated greatly due to several factors, including unprecedented fluctuations in rainfall. However, since 2010, soybean production has become increasingly important (Figure 11.4), and, by 2014, the crop had become the major export item of Roraima, with primary consumer markets in the Netherlands and Russia.

Presently, soybean cultivation takes place in the savanna region, not in the Amazonian tropical rainforest. Production is concentrated around the capital city of Boa Vista, which enjoys better transport infrastructure and many urban amenities (Gianluppi 2008; Gianluppi and Waquil 2008).

Given the specificities of the local climate, the planting period extends from April to May, coinciding with the rainy season, and harvest usually occurs in September. This is a major advantage, because Roraima's soybeans reach the market when the other producing regions of Brazil are off-season and when markets have limited supplies and prices are rising.

In addition to its advantageous seasonal markets, a combination of factors favors Roraima's soybean expansion: much lower land prices, compared to southern Amazonia and central Brazil; a favorable climate; a strategic geographical position and relatively suitable transport logistics. Roraima's

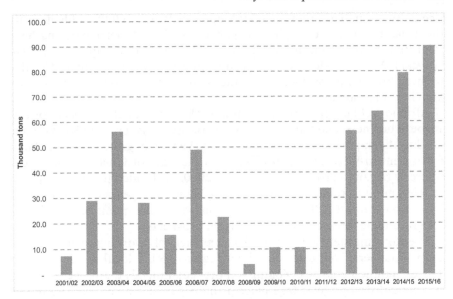

Figure 11.4 The evolution of soybean production in Roraima.

Source: Companhia Nacional de Abastecimento (CONAB), 2017

grains can reach the Caribbean using local roads leading to Georgetown, Guyana (550 km from Boa Vista), or Puerto La Cruz, Venezuela (1,200 km from Boa Vista). The port of Itacoatiara, on the banks of the Amazon River, in the Amazonas state (around 1,000 km from Boa Vista), can also be used as an outlet for production. Roraima is physically much closer to several important consumer markets than other Brazilian soybean producing areas (Figure 11.2).

Reenacting the protagonist role played in the expansion of soybeans in central Brazil, EMBRAPA has been instrumental in developing plant species suitable to the soils and climatic conditions of Roraima's savannas. The adapted species have very short and efficient production cycles and can be harvested in 85 days, adding to Roraima's competitiveness.

Despite the growing importance of the crop for Roraima's economy, local soybean outputs are still exceeded by those of the more-established producing areas of Brazil, both within and outside the Amazon region (Table 11.1).

The encroachment of soybean plantations into Roraima is closely related to the historical expansion of the crop in Brazil, which started in the southern states during the 1950s, reaching the Central-West in the 1970s and from there expanding into the southern peripheries of the Amazon in the 1990s. Roraima represents the latest chapter of this process and, according to Gianluppi and Waquil (2008), the local soybean producers are for the most part newcomers to the state, holding less than five years of residency. Another interesting aspect is that some were previously engaged in rice cultivation. The study by Gianluppi and Waquil (2008) adds flesh to the structural bones of soybean expansion in Brazil, demonstrating how migration histories mimic the spatial evolution of the crop over the last 20 years. Many producers were born in South Brazil and engaged in the soybean production business in Southeast and Central-West Brazil prior to arriving in Roraima.

Afranio Vebber epitomizes this trend. He is one of the pioneers of the soybean business, and today is its leading producer. A native to Rio Grande do Sul, the southernmost Brazilian state, Vebber migrated with his family to Central-West Brazil, specifically to the state of Goiás, in the 1970s, where he began to produce soybeans. In 2000, he decided to buy a property in Roraima, where he began to grow soybeans four years later. Presently, Vebber has almost 2,000 ha under production in the municipality of Alto Alegre, and he foresees serious expansion prospects.

The Roraima state government has been actively working to attract soybean producers from other Brazilian regions to the local savanna areas. The Economic Ecological Zoning Report recommends that large commercial agricultural projects take place primarily in savanna areas and in the already deforested tracts of forest, thereby preserving the intact rainforest. Government officials have justified soybean expansion on the basis of its economic advantages, ignoring the challenges that producers encounter and neglecting to take notice of many potential negative impacts.

Among the challenges that producers encounter are the higher costs of agricultural inputs such as herbicides, insecticides and fertilizers and the slow pace at which new technologies reach the state (Albuquerque et al. 2017). Another severe problem, this time not exclusive to soybean producers, is the lack of land titling because most of the areas under cultivation either were granted by federal and state level officials or were informally acquired, given that squatters and land invaders are not unheard of among these players (Gianluppi and Waquil 2008).

The impact of this embryonic industry upon the local land use structure is already noticeable. Satellite imagery indicates a reduction of 146.4 km² in the natural vegetation in central Roraima between 2000 and 2010 accompanied by a considerable expansion of agricultural lands, which grew from 50.9 km² in 2000 to 182.7 km² in 2010. Concomitantly, one observes in recent years an escalation in land concentration, mechanization, the use of pesticides and diminishing investments in peasant agriculture, phenomena previously witnessed in central Brazil during the 1970s (Silva, Oliveira and Nascimento 2015).

Moreover, the intensive cultivation of local soils alters its physical characteristics to varying degrees, leading to erosive processes, compaction and contamination and serious environmental disturbances, with spillover effects on the whole landscape. Several waterways in central Roraima have their sources in palm swamps, a highly vulnerable subsystem. Because soybean plantations require considerable amounts of irrigation and the heavy use of fertilizers and pesticides, these volatile natural features have been seriously jeopardized (Vale Júnior and Schaefer 2010).

Soybeans have also raised the potential for agrarian conflicts because Roraima has a high proportion of Indigenous peoples and demarcated native population lands with environmental features highly suitable for the plantations. Furthermore, one cannot lose sight of the effects of land concentration. Roraima contains a large number of rural settlements administered by INCRA that support tens of thousands of active peasants. These agricultural colonies function around family labor and are predominantly subsistence economies. They exhibit severe infrastructural and financial challenges, leading to low levels of colonist attrition. Incoming soybean plantation developers may seek the illegal commercialization of these plots, if not the expulsion of the colonists and the consolidation of large tracts of land under commercial agricultural units. Either forcibly or willingly, peasants impacted in this way end up migrating elsewhere in Roraima, taking advantage of any open public lands still available, or further advancing the agricultural frontier in the forest areas of southern Roraima, thereby exacerbating the state's environmental and social problems.

Finally, as new areas are incorporated into soybean production, other important food crops, such as cereals, vegetables and fruits, are inevitably displaced, potentially causing supply shortages and higher food prices for the local population.

Conclusion

Despite being deeply rooted within the Amazon River basin, Roraima has its share of peculiarities, notably its remoteness and special physical endowments. Given the presence of savannas in its northern fringes, cattle were introduced there much earlier than in other areas of the Amazon and cows outnumbered human beings until relatively recently. Its demographic expansion only became viable after the construction of road links in the late 1970s, and the state continues to receive landless peasants in search of subsistence on the open public lands.

This much later timing of occupation has had important implications for the process of frontier expansion because the environmentally blind nature by which other Amazonian areas were explored and exploited no longer applied when Roraima was opened to the world. As a result, local occupation and exploration has been executed in a more careful fashion, given the watchful eyes of world public opinion and the work of environmental and human rights groups. We reemphasize that a large portion of Roraima's territory still lies within Indigenous reservations and environmental sanctuaries.

Until the soybean plantations arrived, Roraima also differed from other areas of the Amazon given the limited presence of corporate capital.

Nonetheless, expanding international grain markets, coupled with its local economic advantages and geographical position, make Roraima's soybean industry highly competitive in global terms, increasing the prospects of more aggressive corporate actions within the near future. This, in turn, could significantly rearrange power relations among the various actors present in Roraima, a state whose frontier has hitherto been a convergence point for Indigenous peoples, loggers, peasants, ranchers, *garimpeiros*, environmentalist groups, religious denominations and, most recently, soybean farmers.

The recent and expansive growth of soybean plantations in Roraima deserves further attention. The economic gains are undeniable, but their social and environmental impacts are still little investigated, and their effects will only become more evident as time progresses.

The official rhetoric in favor of soybean expansion in the savanna region as a means of preserving the forested areas of southern Roraima, while inducing economic development, is misleading. After all, Roraima's savannas are also home to numerous plant and animal species, displaying a delicate environmental balance, whose complexities are yet to be fully understood by science (Carneiro Filho 1993; Maslin et al. 2012).

Local authorities have had the opportunity to learn many lessons accumulated over the past decades about the pros and cons of soybean plantation development from examples in Central-West Brazil and other Amazonian areas. It is important to take these into consideration as a means of promoting sustainable social and economic development in Roraima.

Notes

1. Large rural properties marked by low productive techniques, quasi-feudalistic working relations and dominated by the rural oligarchy (Graziano 1996).
2. Those who meet most of their subsistence needs but often participate in a minor way in the market economy by selling agricultural surpluses (Wharton 1969).
3. Those engaged in *garimpo* mining.
4. Clandestine mines.
5. The semi-sedentary Yanomamis live in an extensive area in the western Roraima state and in southern Venezuela and move freely across the international border, which presented a particular geopolitical challenge in the 1980s (Mattos 1990; Alen 1992; Foresta 1992).
6. Under the Secretariat of Strategic Affairs, the US$1.4 billion endeavor had a dual purpose. The SIPAM (Amazon's protection system) sought to integrate, evaluate and diffuse information, allowing the design of integrated actions among some 50 civilian governmental institutions taking part in the project. SIVAM (Amazon's vigilance system), on the other hand, collected information about the entire region, easing the development of integrative policies and avoiding overlaps in investments and planning. In order to materialize such a Herculean task, SIVAM used a wide range of sensors such as stationary radar, satellites and geophysical monitors. But SIVAM was also charged with the responsibility of watching over the aerial and terrestrial space of the Brazilian Amazon (Leite, 2002).

References

Albert, B 1992, "The environment, the military and the Yanomami", *Development and Change*, Vol. 23, pp. 35–70.

Albuquerque, J, Santos, T, Castro, T, Melo, V and Rocha, P 2017, "Weed incidence after soybean harvest in no-till and conventional tillage crop rotation systems in Roraima's cerrado", *Planta Daninha*, Vol. 35, pp. 1–12.

Allen, E 1992, Calha Norte: Military Development in Amazônia. Development and Change, Vol. 23, pp. 71–99.

Andrade, M 1989, *Geopolítica do Brasil*, Ática, São Paulo.

Arruda, H 1978, *Colonização oficial e particular*, INCRA, Brasília.

Barros N 1995, *Roraima, Paisagens e Tempo na Amazônia Setentrional*, Editora Universitária, Recife.

Barros, P, Padula, R and Severo, L 2011, "A integração Brasil-Venezuela e o eixo Amazônia-Orinoco", *Boletim de Economia e Política Internacional*, pp. 33–41.

Becker, B 1989, "Grandes projetos e produção de espaço transnacional: uma nova estratégia do Estado na Amazônia", *Revista Brasileira de Geografia*, Vol. 51, No. 4, pp. 7–20.

Becker, B 1990, *Amazônia*, Ática, São Paulo.

Bertrand, J, Cadier, C and Gasquès, J 2005, "O crédito: fator essencial à expansão da soja em Mato Grosso", *Cadernos de Ciência and Tecnologia*, Vol. 22, No. 1, pp. 109–23.

Brown, J, Koeppe M, Coles B and Price, K 2005, "Soybean production and conversion of tropical forest in the Brazilian Amazon: the case of Vilhena, Rondônia", *Ambio*, Vol. 34, No. 6, pp. 462–9.

Carneiro Filho, A 1993, "Cerrados amazônicos: fósseis vivos? Algumas reflexões", *Rev. IG*, São Paulo, Vol. 14, No. 1, pp. 63–8.

CONAB (Companhia Nacional de Abastecimento) 2017, *Séries históricas*, www.conab.gov.br/conteudos.php?a=1252andt=2andPagina_objcmsconteudos=3#A_objcmsconteudos> (viewed 15 August 2017).

Contreiras, H 1998, "Raytheon mostra sua cara", *Istoé*, Vol. 1505, pp. 6–8.

Correa, R 1987, "A periodização da rede urbana da Amazônia", *Revista Brasileira de Geografia*, Vol. 49, No. 3, pp. 39–56.

Costa, W 1991, *O Estado e as Políticas Territoriais no Brasil*, Contexto, São Paulo.

Diniz, A 2002, *Frontier Evolution and Mobility in Volatile Frontier Settlements of the Brazilian Amazon*, PhD dissertation, Arizona State University, Tempe, AZ.

Diniz, A 2003, "Migração e evolução da fronteira agrícola", *Geografia*, Vol. 28, No. 3, pp. 63–78.

Domingues, M and Bermann, C 2012, "O arco de desflorestamento na Amazônia: da pecuária à soja", *Ambiente and Sociedade*, Vol. 15, No. 2, pp. 1–22.

EMBRAPA (Empresa Brasileira de Pesquisa Agropecuária) 2004, Tecnologias da Produção de Soja na Região Central do Brasil, www.embrapa.br/busca-de-publicacoes/-/publicacao/451526/tecnologias-de-producao-de-soja—regiao-central-do-brasil-2004 (viewed 10 August 2017).

Fearnside, P 1989, "Projetos de colonização na Amazônia Brasileira: objetivos conflitantes e capacidade de suporte humano", *Cadernos de Geociências*, Vol. 2, pp. 7–25.

Fearnside, P 2001, "Soybean cultivation as a threat to the environment in Brazil", *Environmental Conservation*, Vol. 28, No. 1, pp. 23–38.

Foresta, R 1992, "Amazonia and the politics of geopolitics", *The Geographical Review*, Vol. 82, No. 2, pp. 128–42.

Freitas, A 1997, *Geografia e História de Roraima*, Grafima, Manaus.

Furley, P and Mougeot, L 1994, *Perspectives in the Forest Frontier, Settlement and Change in Brazilian Roraima*, Routledge, New York.

Gianluppi, F 2008, *Desenvolvimento sustentável e sojicultura em Roraima: trajetórias antagônicas ou conciliáveis?* Programa de Pós-Graduação em Agronegócios, dissertação (mestrado), Universidade Federal do Rio Grande do Sul, Porto Alegre, Brazil.

Gianluppi, F and Waquil, P 2008, "Desenvolvimento sustentável e sojicultora em Roraima: trajetórias antagônicas ou conciliáveis", *XLVI Congresso da Sociedade Brasileira de Economia, Administração e Sociologia Rural*, 20 to 23 July 2008, Rio Branco, Acre.

Godfrey, B 1992, "Migration to the gold-mining frontier in Brazilian Amazônia", *Geographical Review*, Vol. 82, No. 4, pp. 458–69.

Graziano, F 1996, *Qual Reforma Agrária? Terra, Pobreza e Cidadania*, Geração Editorial, São Paulo.

Hecht, S 1984, "Cattle ranching in Amazônia: political and ecological considerations", in M Schmink and C Wood (Eds), *Frontier Expansion in Amazônia*, University of Florida, Gainesville, pp. 366–400.

Henriques, M 1984, "A política de colonizacão dirigida no Brasil: um estudo de caso, Rondônia", *Revista Brasileira de Geografia*, Vol. 46, No. 3–4, pp. 393–414.

Hirakuri, M and Lazzarotto, J 2014, *O Agronegócio da Soja nos Contextos Mundial e Brasileiro*, EMBRAPA Soja, Londrina.

IBGE 1991, *Geografia do Brasil*, Vol. 3, Região Norte, IBGE, Rio de Janeiro.

IBGE 1992, *Censo Demográfico de 1991*, IBGE, Rio de Janeiro.

INPE 2018, Taxas Anuais de Desmatamento na Amazônia Legal Brasileira (AMZ), www.obt.inpe.br/prodes/dashboard/prodes-rates.html (viewed 25 June 2018).

Leite, RC 2002, Sivam: uma oportunidade perdida. Estudos Avancados, Vol. 16, No. 46, pp. 123–130.

MacMillan, G 1995, At the End of the Rainbow? Gold, Land and People in the Brazilian Amazon, Earthscan Publications Ltd, London.

Magalhães, D 1986, *Roraima, Informações Históricas*, Graphos, Rio de Janeiro.

Mahar, D 1979, *Frontier Development Policy in Brazil: A Study of Amazonia*, Praeger, New York.

Maia, A 2013, "Estrutura de ocupações e distribuição de rendimentos: uma análise da experiência Brasileira nos anos 2000", *Revista de Economia Contemporânea*, Vol. 17, No. 2, pp. 276–301.

Maslin, M, Ettwein, V, Boot, C, Bendle, J and Pancost, R 2012, "Amazon fan biomarker evidence against the Pleistocene rainforest refuge hypothesis?", *Journal of Quaternary Science*, Vol. 27, No. 5, pp. 451–60.

Mattei, L 2013, "A reforma agrária Brasileira: evolução do número de famílias assentadas no período pós-redemocratização do país", *Estudos Sociedade e Agricultura*, Vol. 2.

Mattos, M 1975, *Brasil, Geopolitica e Destino*, José Olympio Editora, Rio de Janeiro.

Mattos, M 1980, *Uma Geopolitica Pan-Amazônica*, José Olympio Editora, Rio de Janeiro.

Mattos, MC 1990, Geopolítica e Teoria de Fronteiras, Fronteiras do Brasil, Biblioteca do Exército Editora, Rio de Janeiro.

Melo, S, Rocha, J, Manabe, V, Cervi, W and LamparellI, R 2015, "Expansão do cultivo da soja (Glycine max [L.] Merrill) no Cerrado Brasileiro, por meio de séries temporais de dados MODIS", *Anais XVII Simpósio Brasileiro de Sensoriamento Remoto – SBSR*, João Pessoa-PB, Brasil, 25–29 April 2015, Instituto Nacional de Pesquisas Espaciais.

Missão, M 2006, "Soja: origem, classificação, utilização e uma visão abrangente do mercado", *Maringá Management: Revista de Ciências Empresariais*, Vol. 3, No.1, pp. 7–15.

Oliveira, A 1983, *Ocupacao Humana. Em Amazônia Desenvolvimento, Integração e Ecologia*, Brasiliense, São Paulo.

Oliveira, A 2015, "A Amazônia e a nova geografia da produção da soja", *Terra Livre*, Vol. 1, No. 26, pp. 13–43.

Pinto, M 1997, *Relatório Final da Comissão Especial Mista Destinada a Reavaliar o Projeto Calha Norte*, Congresso Nacional, Brasília.

Pochmann, M 2011, "Políticas sociais e padrão de mudanças no Brasil durante o governo Lula", *Revista SER Social*, Vol. 13, No. 28, pp. 12–40.

Quintslr, S, Bohrer, C and Irving, M 2011, "Políticas públicas para a Amazônia: práticas e representações em disputa", *RDE-Revista de Desenvolvimento Econômico*, Vol. 13, No. 23.

Reis, A 1968, "Porque a Amazônia deve ser Brasileira", *Revista Brasileira de Política Internacional*, Vol. 11, Nos. 41–2.

Ribeiro, L 2013, *Transformações na Ordem Urbana das Metrópoles Brasileiras: 1980/2010 – Hipóteses e Estratégia Teórico-Metodológica para Estudo Comparativo*, Observatório das Metrópoles Instituto Nacional de Ciência e Tecnologia, FAPERJ/CAPES/CNPq, Rio de Janeiro.

Ribeiro, M 2016, "Estrutura social e desigualdade de renda: uma comparação entre os municípios metropolitanos e os não metropolitanos do Brasil entre 2000 e 2010", *Revista brasileira de Estudos Populacionais*, Vol. 33, No. 2, pp. 237–56.

Santos, R and Diniz, A 2004, "Impactos sócio-ambientais na fronteira agrícola de Roraima", *XIV Encontro Nacional de Estudos Populacionais – ABEP*, Caxambú, Minas Gerais, 20–24 September 2004.

Sawyer, D 1981, Ocupação e desocupação da fronteira agrícola no Brasil: ensaio de interpretação estrutural e espacial, *Seminário Sobre Expansão da Fronteira Agropecuária e Meio-Ambiente na América Latina*, Brasília, 10–13 November 1981.

Schilling, R 1981, *O Expansionismo Brasileiro: A Geopolitica do General Golbery e a Diplomacia do Itamarati*, Global, São Paulo.

Serra, M and Fernández, R 2004, "Perspectivas de desenvolvimento da Amazônia: motivos para o otimismo e para o pessimismo", *Economia e Sociedade*, Vol. 13, No. 2 (23), pp. 107–31.

Silva, G, Oliveira, I and Nascimento, D 2015, "Dinâmica multitemporal do uso e cobertura da terra em áreas de savanas no município de Boa Vista-RR (2000/2014)", *XVII Simpósio Brasileiro de Sensoriamento Remoto – SBSR*, João Pessoa-PB, Brasil, 25–29 April 2015.

Silva, R 2015, "Amazônia globalizada: da fronteira agrícola ao território do agronegócio: o exemplo de Rondônia", *Confins*, No. 23.

Silveira, I and Gatti, M 1988, "Notas sobre a ocupação de Roraima, migração e colonização", *Boletin do Museo Paraense Emilio Goeldi: Antropologia*, Vol. 4, No.1, pp. 43–64.

Soler, L, Verburg, PH and Alves, D 2014, "Evolution of land use in the Brazilian Amazon: from frontier expansion to market chain dynamics", *Land*, Vol. 3, pp. 981–1014.

Souza, A 1986, *Roraima, Fatos e Lendas*, Governo do Território Federal de Roraima, Boa Vista.

Stone, R 1985, *Dreams of Amazônia*, Viking, New York.

Superti, E 2012, "A fronteira setentrional da Amazônia Brasileira no contexto das políticas de integração Sul Americana", *PRACS: Revista Eletrônica de Humanidades do Curso de Ciências Sociais da UNIFAP*, Vol. 4, No. 4, pp. 1–16.

Vale Júnior, J and Schaefer, C 2010, *Solos sob savanas de Roraima: gênese, classificação e relações ambientais*, Ióris, Boa Vista.

Valverde, O 1990, "A devastação da floresta Amazônica", *Revista Brasileira de Geografia*, Vol. 52, No. 3, pp. 11–24.

Vera-Diaz, M, Kaufmann, R and Nepstad, D 2009, "The environmental impacts of soybean expansion and infrastructure development in Brazil's Amazon basin", *Global Development and Environment Institute*, Working Paper No. 09-05, Tufts University, Boston, pp. 1–25.

Vieira, E 1971, *Exploração de Diamantes em Roraima: 1939–1970*, Publicação Especial #3, UFSM – Campus Avançado de Roraima, Santa Maria (R.S.).

Wharton, C 1969, *Subsistence Agriculture and Economic Development*, Aldine, Chicago.

12 Conclusion: relatively recent rural settlement schemes – representations, realities and results

Roy Jones
Alexandre M. A. Diniz

Introduction

The case studies presented in this volume exemplify many aspects of the frontier literature discussed in Chapter Two. Frontiers were and, in some contexts still are, turbulent and transitional areas situated at the margins, as Meinig (1962) puts it, of more established settled areas and beyond which lay indigenous populations and pristine lands. Within this context, settlers, regardless of the nature of the settlement schemes or the individual choices that brought them there, were and are confronted with isolated and inhospitable physical environments and lack the most fundamental and basic of services. At and beyond frontiers, settlers enter into new lands and bring these areas and their inhabitants within the domain of wider economies, societies and polities. The historical contexts and the world circumstances in which these processes occur may vary, as do the agencies and agents involved in these expansion processes; nonetheless, the frontier settler experiences depicted here all demonstrate that schemes of agricultural expansion at and beyond frontiers are inevitably uncertain ventures. As such, they require effective promotion, careful planning and implementation and, certainly in the dynamic twentieth century, the capacity to adapt to frequently and rapidly changing circumstances. With these requirements in mind, we now turn to a consideration of the representations, the realities and the results of the settlement schemes described in this volume.

Representations

The case studies presented here demonstrate that, across the twentieth century, land settlement schemes were presented to the participants who involved themselves therein in a very positive light. In several cases, this involved the presence of a "push" factor in migration terms (Lee 1966), a shortcoming in the settlers' personal circumstances at their point of origin, such as persecution of Jews in Eastern Europe in the early twentieth century, unemployment in Britain following World War I or poverty and landlessness in Northeast Brazil later in the century. Push factors were also present in the motivations of many of the schemes' government and private progenitors. Not only were governments in Britain and Brazil seeking ways to alleviate the challenges of unemployment and poverty that they faced, but charitable organisations like the Jewish Colonization Association and entrepreneurs like Henry Ford saw land settlement schemes as a means of alleviating local problems as diverse as ethnic discrimination and lack of supply chain control, respectively. The responsibility that the New Zealand government felt towards its returning service personnel or that the Western Australian government felt towards those who had flocked to the state in the relatively short-lived gold rush could also be said to have pushed them into establishing and supporting the Soldier Settlement Scheme and the Homesteading Act.

Nevertheless, it was the representation of the "pull" factors – the apparently positive characteristics of the schemes and of the lands in which they were located – that could be seen as at least equally, if not more, important in convincing settlers, governments, businesses and organisations to participate in these land development schemes. The prospect of both personal and organisational financial gain was, axiomatically, an important factor. The vast majority of the individual and familial participants in these schemes were people of modest means seeking to better themselves economically. This motivation was replicated on a larger scale by, for example, the Hudson's Bay Company, which had land to dispose of, the Ford Company, which required raw materials for its manufacturing processes and the corporations seeking to profit from the expansion of oil palm production in Papua New Guinea and soy bean production in Brazil in an increasingly globalised world.

These anticipated economic benefits of the land settlement schemes were frequently complemented by their purported political, even geopolitical, benefits for states and governments and their social and personal benefits for the settler families. The Brazilian government saw agricultural development in the Amazon as one means of securing its borders in remote and, as the government saw it, underpopulated parts of its domain, and the Homestead and Dominion Lands Acts sought to secure the American and Canadian governments' holds over newly acquired portions of their national territories. For the individual settlers, virtually all of these land development schemes were presented, not only as economic opportunities but also as sources of personal fulfilment, which could be attained through the autonomy and security and even the prestige that, it was characteristically contended, could be provided by land ownership on the various settlement frontiers. In many cases, economic and social representations of the benefits of the schemes were conflated with discourses of modernity and development. This was most apparent in relation to the oil palm projects in New Britain, where settlers were presented with a vision of a lifestyle comparable to that of a developed country if they moved not only to new land but also to a commercial, rather than a subsistence, mode of production. Fordlandia offered a vision of modern America transplanted to frontier

Brazil, and the reclamation of the polders was presented as an engineering wonder of the day. For many of the other schemes, representations of a new life in a new country, such as Canada or Australia, or even in a new region, such as Amazonia or the Dutch polders, also conflated economic development, technological advancement and social transformation.

A mechanism by which land settlement schemes could, deliberately or coincidentally, be made to appear desirable was the adoption of a selection process for the allocation of land to settlers. The procedures adopted by the Dutch government were both the most comprehensive and the most relevant in agricultural terms. Successful applicants were required to demonstrate farming experience, financial competence, commonality of familial purpose and, for those aspiring to larger landholdings, even "pioneer spirit". Agricultural experience was also a criterion of selection for the soldier settlers in New Zealand. But against this, they were also required to have undergone the traumas of World War I, which may not have been the best mental and emotional preparation for a relatively isolated and challenging life on North Island's rugged forest frontier. The British settlers selected to move to Western Australia or Alberta might have been, in general, healthy, of good character and not too dissimilar to their host communities in cultural terms, but they were largely from urban backgrounds and most lacked any agricultural experience in Britain, let alone in remote areas of dense forest or climatically harsh prairies.

Overall, it was not merely a representation, but also a belief, of most of the individuals, governments, businesses and other organisations that the settlement schemes documented here offered a win-win opportunity. At the settlers' points of origin, the schemes were seen to offer an alleviation of unemployment, poverty, land shortage and/or persecution, not to mention the potential for an easing of the causes of political dissatisfaction and even social unrest from the perspective of governments and the prospect of financial profits for private investors. At their points of destination, it was expected that the settlers would make personal economic progress, establish thriving communities and develop as individuals in a brave new world. The schemes were also seen as aiding host governments in opening up and securing underexploited territory, providing further opportunities for local businesses and facilitating modernisation. These glowing representations of what the settlement schemes might deliver for their participants placed a considerable burden upon the progenitors of and the participants in these schemes to deliver in both the short (the schemes' realities or their lives) and the long term (their results and/or their legacies).

Realities

The first reality facing the settlers was the physical nature of the new environments in which they found themselves, and it is here that the value of Geddes' (1915) dictum of "survey before plan" is most apparent. In most cases, governments had, or were expected to have, conducted land assessments before plots were allocated to the settlers to ensure their suitability for the intended agricultural activity. In several cases, these surveys were clearly deficient. In New Zealand, much of the allocated land was too remote, steep and prone to erosion to permit profitable farming, a fact that was soon acknowledged by the national government when the land valuations were steeply revised downwards. In locations from Australia to Brazil, many of the settler plots were located in areas of poor soil, a problem often compounded in Australia by poor drainage and, in Brazil, by the vulnerability of plantation crops to pests and diseases. Even where poorer land was not allocated, as at Vermilion, Alberta, climatic vicissitudes could occur. Not only was the first cropping year there affected by a severe drought and destructive hailstorms but, since the area was seen as largely hail-free, the authorities had not taken out hail insurance. As the century wore on, the sponsors of at least some of the schemes appear to have taken more care over land selection. Land and infrastructure on the Noordoostpolder were thoroughly prepared before settlement took place and – even though their original inhabitants had to be removed from them – the oil palm plots in New Britain were fit for the intended purpose.

The economic environments experienced by the settlers also varied widely. The push factors of both large-scale unemployment and the desire to repay the "debt of honour" to returned servicemen, prompted the pursuit of land settlement schemes in Australia, Canada and New Zealand in the 1920s. Participating settlers entered into arrangements whereby they were provided with land, and sometimes materials, at reduced rates, thereby incurring a debt to be repaid when the farms became established. However, world events overtook them. The Depression of the 1930s caused massive falls in the value of their produce, rendering debt repayment impossible in many cases and leading to high levels of farm abandonment (Bolton 1994). By contrast, settlement schemes, such as those on the Dutch polders and in New Britain, that were instigated in the period following World War II benefitted from the three decades of the post-war "long boom" and were able to progress more successfully through their vulnerable early years. The late twentieth century has witnessed the widespread adoption of neoliberal economic policies and the development of global – but often global core controlled – transport networks for agricultural produce. While these have certainly contributed to the recent expansion of soybean production in Brazil's more remote provinces, they also render them more vulnerable to price fluctuations of the kind that adversely affected New Britain's oil palm plantations in recent decades.

To complete the sustainability triad, it is necessary to consider the social environments of the settlers on these schemes. Here it is appropriate to distinguish between those settlers who were operating in familiar physical, economic and/or social environments and those who were not. At one extreme were the stringently selected farmers on the Noordoostpolder who were settled within a carefully replicated microcosm of their own country and were expected to farm in a manner that equated with the best practice that many of them had already been adopting on farms at no great distance from their new homes. Where the settlers were less familiar, either with the type of farming expected of them (or even of any form of farming at all) or with the country and culture to which they had moved, it would be reasonable to expect that the sponsors of the schemes would supply greater levels of help and assistance in the schemes' early years. This was certainly the case in New Britain where high levels of advice were initially offered to formerly subsistence farmers who had moved to a linguistically different area where they needed to rapidly adapt to the constraints and conditions of commercial farming.

Elsewhere, this was not necessarily the case. Even though they were intranational migrants, soldier settlers in New Zealand were initially blamed for their own failures; and the experience of settler Jim Shaw at least would suggest that government oversight of the settlers was punitive rather than supportive. For those settlers who had moved far from their homelands, such support was even more important, yet the technical and educational assistance provided to the group settlers in Western Australia was notoriously inadequate. However, it was the settlers sponsored by the Jewish Colonization Association on the Canadian Prairies who appear to have experienced the most extreme social problems. The dispersed patterns of rural settlement dictated by the Dominion Lands Act were totally at odds with the requirements of their culture and religion, which were based on close-knit communities within which rituals could be performed with minimal recourse to travel, strict dietary patterns could be followed and within which group marriage was feasible. The Colonization Association largely adopted a "hands-off" approach once initial financial support had been provided to the settlers. For migrants whose cultures and environmental and work experience were compatible with prairie life, such as Christian farmers from the Ukrainian steppes, such a long-distance transplantation was possible, but this was not the case for the largely urban Jewish settlers on the prairies.

One generalisation that can be made about these detailed studies of the quality of life, in their early stages of development, on this diverse set of land settlement schemes that vary so widely across time and space, is that their fates were, to a large extent, dictated by factors beyond, and often far beyond, the settlers' control. Environmentally, they were dependant on those allocating the land to make suitable selections for them. Economically, they were dependant on the sponsors of the schemes to provide levels of guidance and assistance concomitant to the settlers' levels of expertise and financial independence. Furthermore, both the settlers and their sponsors were at the financial mercy of global markets and of events elsewhere (such as the successful development of rubber plantations in Southeast Asia) over the course of a century that was both highly eventful and increasingly dynamic. Socially, they often found themselves in new lands where either the cultural mores or the government regulations were incompatible with the ways of life with which they were familiar. In these circumstances, it is not surprising that the legacies of these schemes are highly variable.

Results

The massive and, if anything, accelerating technological changes that have occurred over the twentieth century have probably had the greatest impacts on the legacies of these schemes. In the late nineteenth and early twentieth centuries, the idea and the ideal of the small family farm could still have currency. In devising their Homesteading Acts, governments in the United States, Canada and Australia could still conceive of a "bold yeomanry", to use John Forrest's term (Tonts 2002), who could aspire to live as a nuclear family unit and make a living on small properties that were allocated to them on what was then seen as virgin land. This ideal survived in various ways throughout most of the schemes depicted here. The Ijsselmeer polders were originally seen as a landscape of small farms, mirroring in size and purpose those established on the "mainland" in earlier centuries. The New Britain plantation scheme allocated small family plots to their oil palm producers. Many, if not most, of the Brazilian government schemes for development in the Amazon in the later part of the twentieth century sought to provide opportunities for small farmers to develop "new" land. By the mid-twentieth century, however, a combination of ongoing technological change and incipient economic globalisation were bringing about agricultural mechanisation and farm amalgamation, as well as the development of global supply systems that had the power to magnify commodity price fluctuations – processes that had the potential to render the small family farm obsolete in reality, if not as an ideal.

Even where agricultural production of a type not too dissimilar to that envisaged in the original schemes continues on these recently settled lands, as is the case in the Canadian prairies, the American Midwest and the West Australian Wheatbelt, far fewer people are now needed for this purpose. In these regions, the legacy of the schemes described in this volume, and of many others, remains in terms of their continuing and considerable farm output. But their farm populations have declined massively, and many small communities struggle or have disappeared completely as many of the descendants of the "pioneers" seek alternative urban employment. Where such urban employment is less readily available, as in parts of the Amazon far from towns or in New Britain, the frequently growing population remains on the settlement scheme properties but, in many cases, they are forced to revert to semi-subsistence agriculture. In both eventualities, processes of change have diluted the ideals of development under which these settlement schemes were first set up. Only in the Dutch polders does there seem to have been, from the start, a conscious awareness of the need for any twentieth century settlement scheme both to expect and to plan for ongoing change.

One aspect of this change, which is apparent in several of the case studies, is the shift towards the idea of a post-productive countryside (Almstedt et al. 2014). The first moves towards the protection of natural environments were made in the late nineteenth century, and these accelerated in the twentieth century, even leading to the dispossession of frontier settlers in some cases (Jacoby 2014). This move for the protection of rural environments was complemented, as the twentieth century progressed, by the increasing popularity (and profitability) of the countryside as a site of consumption by tourists, long-distance commuters, retirees and second-home owners. While the case study of the forests of the Australian South West provide the clearest example of such a progression because the subdivision and subsequent abandonment of many small plots facilitated their eventual purchase by those seeking a rural idyll in the forest, the economically marginal nature of the land in several of the case study sites from New Zealand to Brazil has contributed to their redesignation as conservation and/or recreational areas. The Dutch polders provide a particularly interesting example here. The earliest polders were devoted entirely to productivist agricultural pursuits. But, as the twentieth century progressed, increasing proportions of the more recent polders were given over to conservational and recreational activities, an acknowledgement, perhaps, that sustainable land settlement schemes should have positive environmental and social outcomes as well as economic ones.

Conclusion

The schemes described here were both accurately and inaccurately portrayed to their participants. They have been both well and poorly planned and administered. And they have been both successful and unsuccessful in adapting to changing economic, political and social circumstances. That some of the more recent schemes have – so far – seemed to fare better than many of the earlier ones may be ascribed to the overall accumulation of knowledge and skills in the fields of agriculture and remote regional development. However, it is the study of late nineteenth century homesteading legislation that provides the only explicit example of international knowledge transfer in these studies, and the importance of external events, such as the 1930s' Depression, in determining a scheme's success or failure has already been emphasised.

Today, increases in agricultural production are largely being attained through intensification and technological advances, rather than through the settlement of new lands; and the massive expansion of the world's agricultural area in the twentieth century can be contrasted with the shift of the world's population, over the same period, towards an urban majority. The remains of schemes such as Fordlandia and the Group Settlements have become tourist attractions, and the polders have been suggested for a World Heritage listing. The areal extension of our agricultural frontiers may not quite be a thing of the past, but the case studies presented here provide insights into the challenging circumstances under which their last great expansion was achieved.

References

Almstedt, A, Brouder, P, Karlsson, S and Lundmark, L 2014, "Beyond post-productivism; from rural policy discourse to rural diversity" *European Countryside*, Vol. 4, pp. 297–306.

Bolton, G 1994, *A Fine Country to Starve in*, University of Western AustraliaPress, Nedlands.

Geddes, P 1915, *Cities in Evolution*, Williams and Norgate, London.

Jacoby, K 2014 *Crimes against Nature: Squatters, Poachers, Thieves and the Hidden History of American Conservation*, University of California, Berkeley.

Lee, ES 1966, "A theory of migration", *Demography*, Vol. 3, No. 1, pp. 47–57.

Meinig, D 1962, *On the Margins of the Good Earth: The South Australian Wheat Frontier 1869–1884*, Rigby, Adelaide.

Tonts, M 2002, "State policy and the yeoman ideal", *Landscape Research*, Vol. 27, No.1, pp. 103–15.

Index

Milton Keynes UK
Ingram Content Group UK Ltd.
UKHW040107071024
449327UK00019B/864